Site Investigation and the Law

Jack Cottington
Robert Akenhead

Thomas Telford Ltd
London, 1984

Throughout the text the *ICE Conditions of Contract for Ground Investigation* is referred to as the *Conditions*. The *Conditions of Contract for Works of Civil Engineering Construction* is referred to as the *ICE Fifth Edition*.
Terms used in this book include terms which are defined in the *Conditions*. These terms are printed with capital letters to indicate that the defined meaning is intended. Clause numbers and paragraph numbers refer to those in the *Conditions* unless otherwise stated. Italics in extracts from the *Conditions* are authors' emphasis.

Published by Thomas Telford Ltd
PO Box 101
26–34 Old Street
London EC1P 1JH

British Library Cataloguing in Publication Data

    Cottington, John
       Site investigation and the law.
       1. Building sites – Great Britain
       2. Building – Contracts and specifications –
      Great Britain
       I. Title   II. Akenhead, Robert
       624   TH375

ISBN: 0 7277 0188 6

Set in Bembo by Katerprint Co. Ltd, Oxford.
Printed and bound in Great Britain by
Redwood Burn Limited, Trowbridge, Wiltshire

Contents

Foreword

by Leslie G. Deuce, FEng, FICE, FIHE
Chairman, Joint Committee to draft Conditions of Contract
for Ground Investigation

I am very pleased that this book has been written and I have great pleasure in introducing it to anyone who has an involvement in the commissioning or conduct of ground investigation contracts. It is designed as a guide to the *ICE Conditions of Contract for Ground Investigation* but it is more than that. It contains an analysis of the clauses, advice upon their interpretation and use, and sets them in the context of a clear understanding of the law and the practice of ground investigation. It is published to coincide with the introduction of the new form and I believe it will prove to be a valuable aid to the understanding and acceptance of the form.

I do not suppose for one moment that the form will be without criticism but I am confident that if widely adopted the form has the potential to improve significantly the conduct, practice and efficiency of ground investigation contracts. Based upon the *ICE Fifth Edition*, it is designed to take account of the considerable differences between a civil engineering contract, in which the end product is some form of physical entity, and the scientific and exploratory nature of ground investigation work, which, in contrast, will ususaly aim to leave the site in an apparently undisturbed state.

There will be some who will feel that a much simpler form should have been written. I would not dispute the desirability of this but I would argue that the use that has been made of the tried and trusted clauses of the *ICE Fifth Edition*, its familiar clause numbers and its familiar layout will enable the new form to be readily assimilated by engineers and contractors alike once they have grasped and understood the reasons for the differences. In implementing this approach the drafting committee steadfastly avoided making changes which were not absolutely necessary for their purpose and thus hoped to avoid changing the balance of risk between the parties and jeopardising the interpretation of either of the forms.

This book, written in partnership by a member of the Joint ICE, ACE, FCEC Committee which drafted the Conditions and a Barrister who specialises in contract law, should do much to promote an early understanding and thereby provide a service of considerable value to all sides of the industry.

Leslie G. Deuce
Sevenoaks
October 1983

Preface

The cure for this ill is not to sit still,
Or frowst with a book by the fire,
But to take a large hoe and a shovel also,
And dig till you gently perspire.
Rudyard Kipling

The science of ground investigation has progressed since the days of Rudyard Kipling and in 1977 a working party of the Institution of Civil Engineers Engineering Committee recommended that special contract conditions were necessary for ground investigation.

This guide is written for many 'conditions of men'. Some will be fully experienced and qualified in the art of site investigation and foundation design. Others will be estate agents, potential developers, either private companies or local authorities and their engineers or architects, who consider that purchasing a particular site may be a good investment. And, finally, the guide is written for the Contractor upon whose skill so much depends.

There is no apology to those in the first category for presenting very basic considerations in the first chapter. It may stir them, in the academic or administrative realms in which they probably now dwell, to consider, once again, the primary requirements of site investigation. In later chapters we provide guidance from basic concept through to contractual responsibility.

The *ICE Conditions of Contract for Ground Investigation* relate to the investigation and its findings which must be presented in a form adequate to design economically for the safe use and stability of the structure and the ground concerned. This is the basic difference between the *Conditions* and the *Conditions of Contract for Works of Civil Engineering Construction*.

The new *Conditions* are based on the premise that there are four normal stages of a ground investigation: the desk study;

Site Operations; laboratory and *in situ* testing; and the Report. The technical aspects relate directly to BS 5930:1981 *Code of practice on Site Investigation*; the British Standard *Code of Practice on the identification and investigation of contaminated land* (in draft); CP 2004: 1972 *Foundations* and the British Standards on soil testing. The *Conditions* follow the format and clause sequence of the *ICE Fifth Edition*.

This book advises on the principal technical applications and the new legal implications of the *Conditions*, with particular reference to new or modified clauses. Examples of case law relating to site and ground investigation based on the *ICE Fifth Edition* are discussed in the light of the new *Conditions*.

These new *Conditions* do not relate only to large investigations for major works, they also apply to small and forensic investigations. The responsibilities of the parties concerned are discussed and also assessment, setting out, execution, making good, *in situ* and laboratory testing, the preparation of reports and provision for insurance. The execution of permanent works, called Ancillary Works in the *Conditions*, is also dealt with. In major civil engineering works these could be stations monitoring areas of potential landslide or making long-term observations of seasonal, cyclic and ground conditions in areas subject to chemical action and contamination.

The authors recommend the *Conditions* for use as the basis of Contract in all cases involving ground investigation, because these *Conditions* recognise and safeguard the various operations whatever the reasons for the investigation. By invoking particular clauses of codes and standards dealing with site investigations, the *Conditions* can give 'teeth' to those codes, making their salient features mandatory on the parties concerned.

Clause 72, 'Special Conditions', and also the Specification can be used for this purpose. The clause must be drafted with legal advice.

A contract based on these *Conditions* requires that both Engineer and Contractor have site experience and a detailed and working knowledge of the investigation codes. Conditions on different sites will vary and this will have to be recognised in the tailoring of the Contract documents for the specific project. The purposes of this book, therefore, are:

(*a*)   to aid the preparation of the Contract documents;
(*b*)   to examine the meaning and application of the Contract;
(*c*)   to indicate the courses open for the resolution of disputes.

The Contract will be successful only if the ingredients are fair to both parties and goodwill is maintained throughout its execution.

In another poem Kipling said,

> *I keep six honest serving-men,*
> *(They taught me all I know);*
> *Their names are What and Why and When*
> *And How and Where and Who.*

In the following pages, we present an answer to those men.

Robert Akenhead
of the Inner Temple
Lincoln's Inn

Jack Cottington
Cottington Phillips & Associates
Bridport
Dorset

Acknowledgements

It is fitting that the Foreword to this work is given by Leslie G. Deuce FICE, without whose chairmanship and personal contribution, the *ICE Conditions of Contract for Ground Investigation* would probably still be in the process of drafting. Other members of the Contract Drafting Committee were Dr. A.C. Meigh, G.B. Britt, J. Cottington, D.H.L. Keeble, H.G. Clapham, F.H. Hughes and the Committee Secretary, H.A. Jones.

The authors have been most fortunate to have the unstinting co-operation, support and encouragement of the several institutions and colleagues whose publications are quoted and referred to herein and form much of the basis of presentation. The support of our professional colleagues and friends, even after reading draft chapters, the understanding of our families during times of stress and the many others who have suffered indirectly, have resulted in what we trust will be something of value.

Our publisher's advice and confidence, particularly that of Jeremy Swinfen Green and Patricia Monahan, has been invaluable, as has the patience of Janet Gilbertson, without whose agile brain and nimble fingers this work may have foundered, which is much appreciated.

We formally thank and acknowledge those named below, who permitted us to use their work.

The Institution of Civil Engineers of Great George Street, London SW1, for permission to quote from and include as Appendix 1 *The ICE Conditions of Contract for Ground Investigation*.

The British Standards Institution of 2 Park Street, London W1A 2BS, for permission to quote from, and refer to, their several codes and standards and from whom complete copies of those publications concerned may be obtained.

The Association of Consulting Engineers of Alliance House, 12 Caxton Street, London SW1 0QL, for permission to refer to *ACE Conditions of Engagement 1981* and from whom complete copies may be obtained.

The Federation of Civil Engineering Contractors of Cowdray House, 6 Portugal Street, London WC2A 2HH, for permission to reproduce the *Schedules of Dayworks Carried out Incidental to Ground Investigation Site Operations* as Appendix 2 and from whom complete copies may be obtained.

Messrs. H.G. Clapham and F.H. Hughes for permission to use material from their article 'Ground Investigation: Schedules of Dayworks' published by Morgan Grampian (Construction Press) Ltd in *Civil Engineering*, April 1983.

Morgan Grampian (Construction Press) Ltd of 30 Calderwood Street, London SE18 6QH, for permission to use material from articles by J. Cottington previously published by them.

J.C., R.A.

Table of cases

The Building Law Report references are given where applicable

Chapter 1
A background to the preparation of Contract Documents

He who would do good to another must
do it in minute Particulars.
General Good is the plea of the scoundrel,
hypocrite and flatterer;
For Art and Science cannot exist but in
minutely organised Particulars.
William Blake

Introduction

It is essential that the parties to a Contract are familiar with terms which have defined meanings. These terms are used throughout the *Conditions* and they are also used in this book.

The clause 1(1) definitions are set out in full in Appendix 1, pp. 126–163. A number of new definitions which may not be familiar in a normal civil engineering contractual context have been introduced into the Conditions. These are set out below:

(*h*)   'Schedules' means the schedules and lists of Site Operations Laboratory Testing and other requirements referred to in the Specification;

(*k*)   'Site Operations' means all the work of every kind including Ancillary Works required to be carried out on under in or through the Site in accordance with the Contract;

(*l*)   'Ancillary Works' means all appliances or things of whatsoever nature required to be installed or constructed on under in or through the Site and which are to remain on Site and become the property of the Employer in accordance with the Contract upon issue of a Certificate of Completion in respect of the Site operations or section or part thereof as the case may be;

(*m*)   'Laboratory Testing' means the testing operations and processes necessary for the preparation of the Report to be carried out in accordance with the Contract at a laboratory approved by the Engineer on samples and cores obtained during the Site Operations;

(*n*)   'Report' means the report to be prepared and submitted in accordance with the Contract;

(*o*)   'Investigation' means the Site Operations, together with the Laboratory Testing and Report preparation and submission.

(*r*)   'Equipment' means any appliances or things of whatsoever nature required temporarily for carrying out the Site Operations but does not include anything which forms part of the Ancillary Works.

The fundamental purpose of this Contract is the execution of the Investigation, and it is for this that these *Conditions* have been prepared. The Investigation may include as Site Operations all temporary or permanent *in situ* works (the latter known as Ancillary Works), as well as all the specified off-site Laboratory Testing and other processes. The practical conclusion of the Investigation is the submission of the approved final Report.

It is emphasised that the *Conditions* as printed are for the execution of a ground investigation; they do not necessarily cater for a site investigation (as envisaged in BS 5930: 1981). If a site investigation is required by the Employer, or the Engineer on his behalf, this must be clearly stated in the Contract Agreement, clause 72 and the Specification.

Interested parties

There are a number of parties whose functions are regulated or provided for in the Conditions of Contract: the Employer, the Engineer, the Engineer's Representative, the Contractor and sub-contractors.

The employer

Recent decisions in the Court of Appeal (Batty v Metropolitan Property Realisations Ltd 1978 OB 554, Acrecrest Ltd v W.S. Hattrell and Partners 1982 3 WLR 1976, Eames v N. Herts District Council 18 BLR 50) have made it clear that the developer (who will usually be the Employer) owes a duty in law (independent of contract) to examine land on which they intend to build, to ascertain whether it is land on which they can safely build.

The consequence of the law as it now stands is that, save in exceptional circumstances, the Employer will take an unacceptable risk if he fails to have an adequate ground investigation carried out. Experienced developers (and engineers) will lay themselves open to charges of actionable negligence if they initiate or recommend inadequate ground investigations.

The Employer, who is to be defined by name in the Conditions of Contract, clause 1(1)(*a*), does not normally warrant, expressly or by implication, that the Investigation is physically capable of being carried out by the Contractor

(Thorn v London Corporation 1876 1 App Cas 120). However, if performance proves impossible, the Engineer is obliged to issue appropriate instructions on behalf of his Employer. (See Chapter 2, p.17.)

The Employer is responsible for ensuring that all appropriate statutory consents (e.g. planning and building regulations approvals) are obtained in sufficient time to enable the Contractor to perform his obligations under the Contract.

The general obligations of the Employer (payments, possession) are considered in detail in Chapter 3.

The *Conditions* render the Employer liable to the Contractor for acts or omissions of the Engineer which are not reasonably foreseeable and cause the Contractor to incur delays and additional costs.

The Employer's experience of design, construction and development may vary from complete ignorance, or superficial expertise, to great detailed knowledge. The Employer may be a government department, a local authority, a financing institution, a large or small developer, a small builder or the unfortunate owner of a failed structure. The less the technical experience of the Employer, the more necessary it is to state clearly the object of the Investigation at Tender stage in the Preamble of the Specification and in the Specification; this will ensure that the Contractor acknowledges his responsibilities at Contract stage.

The Engineer
There are several categories of employment and Engineers will probably find themselves within one of those listed below.

(*a*)   an officer of a government department or establishment;
(*b*)   an officer of a local authority;
(*c*)   an engineer within a multi-discipline or specialist consulting practice;
(*d*)   a self-employed consulting engineer;
(*e*)   an employee engineer to a construction firm;
(*f*)   an employee engineer of a commercial or industrial organisation.

It follows that the scope of operation, degree of responsibility and geotechnical experience will vary considerably.

3

Similarly, the geotechnical resources and experienced technical support may differ from adequate to inadequate.

The Engineer will owe to his client, the Employer, duties in contract and in tort to exercise all reasonable care and skill in the provision of engineering services. If the Engineer recognises that he does not have the expertise to direct, process or use the Investigation, it is in his and his client's interest that this is communicated to his client. The Engineer can then legitimately recommend that a full site investigation be performed by the Contractor, encompassing a ground investigation, together with detailed proposals as to appropriate designs.

The variations in category, experience and support dictate the degree to which the Engineer is able, or needs, to prepare the Contract documents and whether the supervision required will be personal or delegated.

The Engineer in category (*a*) above may always have used standard specifications and documents. Although care should be exercised to avoid ambiguities, the *Conditions* can easily be adapted to incorporate these special documents by amending the Form of Tender, clause 1(1)(*e*) (definition of Contract Documents), and possibly clause 36(1) (quality of workmanship and materials).

It will rarely be in the interests of the client for the Engineer to withhold information from the tenderers at tender stage. Indeed, withholding such information might lead to increases in cost; a result of clause 12 claims at a later stage.

As well as owing duties to the clients, the Engineer and those acting under him owe duties in tort, to exercise reasonable care in the discharge of their functions. The duties are owed to any third party who foreseeably is likely to be injured or to suffer loss as a result of a failure to exercise reasonable care. Thus an Engineer, who negligently instructs the Contractor to carry out work in a dangerous manner, will be liable directly to an individual who is injured as a result; the fact that, between the Employer and the Contractor under the ground Investigation Contract, the Contractor or Contractor's insurers are liable, will not avail the Engineer.

The Engineer's Representative
In general terms only, the function and contractual position of the Engineer's Representative are the same under these Condi-

tions as under the current *ICE Fifth Edition*. In practice, the responsibilities of his office under these *Conditions* are extended.

In a ground investigation Contract, the Engineer's Representative will be required not simply to supervise the Contractor but, more importantly, to monitor and adjust the Investigation as it proceeds. He will have to be sufficiently experienced to recognise critical changes in soil type or conditions which not only affect the Investigation itself but are of relevance to future design. Of particular importance will be the ability to recognise the need for the involvement of specialists to assist the Engineer. The Employer and Engineer can retain specialists by using the machinery of clause 2; this can be achieved by naming such persons as temporary or permanent Engineer's Representative or as assistant under clause 2(2).

The Engineer's Representative and assistants can only act as such if their respective appointments have been notified in writing to the Contractor.

It will be seen that the professional responsibility of these persons may well be increased in the case of ground investigations on sites of dubious history.

The Contractor
The Contractor is defined as being the entity whose tender has been accepted by the Employer. It is not legitimate for the Contractor to assign or sub-let the Contract or the Investigation without the written consent of the Employer or Engineer respectively.

The primary obligation of the Contractor is to carry out and complete the Investigation, by so doing he thus warrants that he has the appropriate expertise and capability. This is recognised by clauses 8 and 16.

The obligations and liabilities of the Contractor are considered in detail in Chapter 2 (et seq).

Sub-contractors
The Contractor is responsible contractually to the Employer for the defaults of his sub-contractors, whether nominated or domestic. In the absence of any direct agreement between a sub-contractor and the Employer, the sub-contractor owes

contractual duties only to the Contractor. The Employer and Engineer can influence the selection of sub-contractors and the agreement of sub-contract terms by two methods: by initial incorporation of proposed sub-contractors' work as a Prime Cost Item (clauses 58 and 59A) and by the judicious use of clause 4.

Relatively recent legal developments have established that sub-contractors owe to Employers, future owners or occupiers and to third parties, a duty in tort (independent of contract) to exercise reasonable care in carrying out their sub-contract functions (e.g. Junior Books Ltd v Veitchi Co Ltd 1982 3 WLR 477, 21 BLR 66).

Objectives

The object of a ground Investigation must be established before attempting to plan the Specification. There may be a series of objectives to be achieved. It is therefore advisable to prepare a plan of operations to be covered by the Contract either in tabular or check-list form. This plan will be related to the several stages covered by the general obligations (clauses 8 to 33).

An example of this is given in Figure 1. The plan of operations will be related to BS 5930: 1981 extended, when necessary, by the relevant codes of practice. These will be involved when the investigation of contaminated land is concerned and the execution of specific Laboratory Testing is required. This relationship may need to be made mandatory under clause 72 (special conditions) or in the Specification and schedule of tests. To ensure that guidance on particular operations is not taken out of context, both the Engineer and the Contractor should study the opening sections of the codes of practice concerned before relating the codes of practice to the detail of a specific operation.

The object of the Investigation may be to provide, or seek for, specific information relating to a proposal, requirement or problem. An instruction requesting this information must be provided, in full narrative form or in précis, and this will be repeated in the Specification. The reason for the Investigation should usually be stated, but sometimes, in the interests of national or commercial security, it is not stated. Even so, the requirements and, as far as possible, the objectives of the Investigation, must be given.

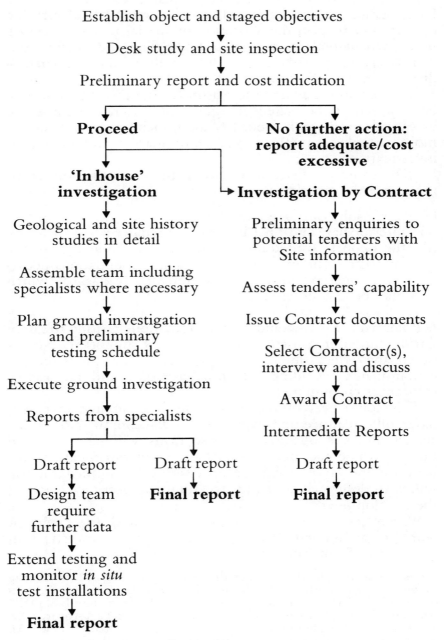

Establish object and staged objectives

Desk study and site inspection

Preliminary report and cost indication

Proceed

No further action: report adequate/cost excessive

'In house' investigation

Investigation by Contract

Geological and site history studies in detail

Preliminary enquiries to potential tenderers with Site information

Assemble team including specialists where necessary

Assess tenderers' capability

Plan ground investigation and preliminary testing schedule

Issue Contract documents

Execute ground investigation

Select Contractor(s), interview and discuss

Reports from specialists

Award Contract

Intermediate Reports

Draft report   Draft report   Draft report

Design team require further data

Final report   **Final report**

Extend testing and monitor *in situ* test installations

Final report

Figure 1.   Programme of operations

If there are no restrictions on publicity, the reason for the Investigation should be stated, thereby giving tenderers the opportunity to plan the allocation of specialist staff, together with particular plant, methods of sampling and *in situ* tests. This is advantageous to economy and will avoid misunderstanding of the overall requirement.

Some examples where the reason for the Investigation may be required to be stated are given here. They assume the investigation of virgin land, but the principles apply to other, more complex situations, including marine and inter-tidal investigations.

State exactly what information the Investigation is to provide, for example:

(a) test results only are required
(b) intermediate and final Reports are required
(c) the Reports are required to include recommendations for land-use, type of drainage, roads and foundations necessary, together with an indication of any hazards which may be encountered.

Potential hazards would include limitations relating to safety during and after construction, in terms of support to excavations, the effect on adjacent properties and the effect of contaminants including inflammable, toxic and radioactive substances encountered.

This could be required where the firms invited to tender have fully qualified and experienced geotechnical staff, laboratories and site operatives. These firms will be capable of meeting these requirements including recommendations for land-use and layout, and the provision of quantified data on which to design all civil and structural engineering aspects of the development.

Where the Employer has no geotechnical experience it would be advisable to present the reason for the investigation in narrative form, such as:

It is proposed to develop a site on the northern outskirts of Brobdingnag for light industrial, office and residential purposes. The site comprises plots (parcels) 7432, 7433, 7434 and 7435 as shown on OS map sheet No. OXO with a total area of 100 acres (40.5 hectares). As far as can be discovered, the area has always been heathland and pasture with the exception of plot 7432, which has been under continuous cultivation for at least 150 years.

Subject to the result of a site investigation, plots 7432 and 7433 are intended for light industry, plot 7434 (currently marshy ground) as a recreational and community area, with the remainder for residential properties. The residential properties envisaged will consist of three-storey flats, two-storey four-unit terraces and groups of single-storey accommodation at an approximate density of 10 to the acre (0.4 hectare).

The investigation is required to determine the engineering properties of the soil and the behaviour of the groundwater on which to base the design and construction of surface, sub-surface and foul drainage, road construction to maximum 5-ton single-wheel load and foundation design for all structures. Information is required on all hazards to health and safety within the site which may be related thereto during and on completion of construction.

It is proposed to construct foul and surface water pumping stations in plot 7434 with rising mains in the general direction indicated.

The internal road layout is yet to be decided but its junction with highway B1234 will be made on its boundary with plot 7433. Formation level is expected to be approximately 3 feet 3 inches (1 metre) below ground level at the junction, falling to approximately 6 feet 6 inches (2m) at the northern end of the site.

The land is owned by the Employer and free access is permissible subject to conditions given in the Specification.

The request for information could also be expressed in the following summarised form:

(a) Site location on print from OS map sheet to scale 1:50000.
(b) Site boundaries on print from contoured OS map sheet to scale 1:2500 giving plot (parcel) numbers and area. Indicate access points.
(c) Geological information and land conditions during inclement seasons.
(d) Site history and present condition re access, previous and existing buildings, and condition of adjacent structures.
(e) Details of proposals commencing with broad outline of layout being considered and continuing with detail of development envisaged.
(f) Anticipated road layout, junctions with highways, traffic loadings and approximate formation level.
(g) Foul, surface and sub-surface drainage anticipated, including approximate invert levels.
(h) Names of landowners and detail of entry permission.
(i) Indication of preferred starting, duration period and completion date.

The fact that the stage has been reached where a ground investigation is to be undertaken shows that a desk study, or at least a preliminary appraisal, has been completed otherwise it would be illogical and expensive to go into contract. Without a desk study or preliminary appraisal it would be unlikely that recommendations based on the result of the ground investigation would be accurate, because other factors affecting the

9

condition and performance of the ground investigated could be overlooked. The extent of the desk study or preliminary appraisal will be determined by the type of problem, together with the extent of the area and soil mass involved.

Preparation of tender documents

The tender documentation will become part of the Contract so it is important that such documents are prepared with great care. Slipshod or imprecise drafting can lead to claims by the Contractor for misrepresentation, or a generous application of normal claim clauses (e.g. clauses 12 and 13).

Although it is possible, and common, to enter into a Contract based on these *Conditions* informally, by letter for example, it is advisable to adopt the contractual procedures recommended within the *Conditions* themselves. The documents which are recognised as comprising the Contract are dealt with separately below.

The tender

A Form of Tender is provided with the standard Conditions of Contract and consists of a standard offer to carry out the Investigation in accordance with the Contract and a detailed Appendix. It is crucial that the Appendix is complete by the time the Tender is accepted, since failure to complete could result in terms being implied into the Contract which either party might not have intended; for instance, failure to state the time for completion could result in the completion date being put at large. (This means that the Contractor is no longer held to that date but to that agreed within which it is reasonable to expect him to complete.)

In the interests of comparable competitive tendering, it is in the Employer's interest at Tender stage to specify his requirements in the Appendix.

Conditions of Contract

Guidance note 2B of the ICE Conditions of Contract Standing Joint Committee recommends that clauses 1 to 71 should not be altered if it is intended to use those *Conditions*. In any event, great care should be exercised (and legal advice taken) if making any alterations to the *Conditions*; for instance, if the parties agree upon a lump sum basis for payment, wholesale modification of clauses 52, 55 to 57 and 60 will be required. If it is intended that there should be no Engineer (as defined), these *Conditions* are not appropriate.

If modifications are made, the parties must check by cross-reference between clauses that such change is consonant with other unchanged clauses.

The Specification

The preparation of a Specification for ground investigation is not as straightforward as for construction works. Specific information is required by the Employer and Engineer from a situation and a material both of which are infinitely variable. It follows that, except for extraordinary circumstances, flexibility (but not looseness) in the drafting of the Specification is desirable.

The Specification (discussed in Chapter 4) should identify in detail the object, objectives and extent of the work and services required, as well as the quality and standards to be complied with. The Specification should contain the list of drawings which are intended to be the Contract Drawings (clause 1(1) (*g*)).

Since it is frequently necessary to require compliance with codes of practice, the Specification is the appropriate document in which reference to the codes is invoked. However, wholesale incorporation of such standards should be the exception, because this can lead to discrepancies between the codes of practice themselves and the Contract requirements. Reference to specific sections or clauses of the codes of practice is essential.

Schedules

This class of Contract Document is peculiar to these *Conditions* and should be referred to in the Specification (clause 10 *h*).

Whilst schedules may not be required as separate Contract Documents if the Specification and/or Bills of Quantities are sufficiently detailed, the documents are used to perform two important functions:

(*a*)   to list in detail and tabulate what the Contractor may be called upon to provide as part of his express contractual requirement.

(*b*)   to identify items of work or services which the Contractor may be called upon to provide should the need arise.

11

The Schedules can be used to identify what is foreseeable and/or included in the Contract performance.

Drawings

Any drawings which are to be Contract Documents must be listed in the Specifications. Drawings can illustrate and detail not only the Ancillary Works but also topographical features of the Site (e.g. level surveys).

Priced Bill of Quantities

In ground investigation, the preparation of the Bill of Quantities is primarily the function of a geotechnical engineer or engineer geologist, assisted, when necessary, by other specialists concerned with the Investigation. Engineers lacking such expertise should obtain specialist advice.

The priced Bill of Quantities represents an agreement on the rates and prices applicable. Because of the nature of the work the quantities involved are likely to be more imprecise under these *Conditions* than under the *ICE Fifth Edition*.

Provision of information to tenderers

Whilst there is no obligation at common law on the Employer or Engineer to disclose any information to the Contractor before the tender, it is, on balance, desirable that all relevant information should be provided to tenderers. This should ensure the submission of comparable tenders and reduce the risk of claims. Many claim clauses make the actual, or presumed, knowledge of the Contractor an important prerequisite of his right of recovery. The information which can sensibly be provided in most cases as a reasonable minimum is discussed on p.8.

Providing incorrect information to tenderers could result in a quantifiable loss to the successful tenderer which could, in turn, result in a liability on the part of the Employer for misrepresentation, and on the part of the Engineer for negligence. Conversely, it would be rare for an Employer to recover a windfall profit made by the Contractor as a result of the provision of incorrect information.

Provision for specialists

There are three methods recognised in the *Conditions* by which the Employer may provide for the use of specialist services or techniques.

Express specification

If the Employer or Engineer wishes to require the Contractor in any event to use, as part of the Investigation, a particular organisation or technique for certain services, it should be clearly laid down in the Specification. Clause 15(3) permits this in part:

If . . . the Contract shall require or the Engineer direct the Contractor to make available *on the Site or elsewhere* the services of suitably qualified persons for description of soils and rocks logging of trial pits execution of geological and geotechnical appraisals other technical and advisory services and the preparation of technical reports *the extent and scope of the service required shall be specified in the Contract.*

The advantages of this approach are clear: the Employer and Engineer know what to expect from the Contractor at the date of the Contract, in terms of extent, likely quality and cost of the specialist services. The principal disadvantage arises where the specialist is either unable or unwilling to proceed: the Contractor's delays and costs may well attract extension of time under clause 44 (a special circumstance) and reimbursement of consequential cost under clauses 7(3) or 13(3) (late instructions resolving the impasse), or clause 52 (variations altering the specified requirement).

It is open to the parties to make the specialist services or technique the subject matter of a Provisional Sum. Such an item would be regulated by clauses 58(1) and 58(7)(*a*), with payment as if it were a variation to be valued under clause 52.

Prime Cost (PC) Items

The Employer and Engineer may decide that it is more appropriate to incorporate highly specialist operations into the Contract under Prime Cost Items. Such techniques as continuous sounding by pressurised apparatus, the detection and measurement of radioactive, toxic, inflammable and other contaminated material, or the use of geophysical exploration

13

methods (seismic, electro-resistivity) may well be sensibly dealt with as a Prime Cost Item.

It will be incumbent upon the Engineer to obtain detailed quotations from specialists whose services are covered by a Prime Cost Item, and thereafter to instruct the Contractor to place orders with such specialists in the normal way. If the specialist is to be responsible for an element of design or specification, this responsibility has to be specified in the Contract as part of the Prime Cost Item, if it is to have any contractual effect (clause 58(3)).

The main benefit of the employment, through the Contractor, of specialists as Nominated Sub-contractors is that there is one chain of accountability for both payment and responsibility. The principal disadvantage is the almost inevitable delay and loss brought about by a late, failed or deficient nomination.

Direct employment

The Contract envisages that the Employer is to be permitted to use directly employed contractors on or near the Site. Clause 31(1) states:

> The Contractor shall in accordance with the requirements of the Engineer afford all reasonable facilities for any other contractors employed by the Employer and their workmen and for the workmen of the Employer and of any other properly authorised authorities or statutory bodies who may be employed in the execution *on or near the Site* of any work not in the Contract or of any contract which the Employer may enter into in connection with or ancillary to the Investigation.

It is possible that the Employer or Engineer will wish directly to employ specialists whose duties will not be performed 'on or near the Site'. Such specialists may include toxicologists, chemical analysts, botanists and nuclear scientists, whose functions involve monitoring or further testing of samples obtained by the Contractor. Since, on one reading, clause 31(1) does not require the Contractor to provide facilities for such directly employed specialists away from the Site, it is important that appropriate provision for this is made expressly in the Specification.

The main advantage of using directly employed specialists is that closer control can be exercised over their performance. The corresponding disadvantage is that, unless their performance is synchronised with that of the Contractor, the latter may well have grounds for claiming extension of time and consequential cost (clauses 44 and 31(2) respectively).

Formation of Contract

A contract based on the *ICE Conditions of Contract for Ground Investigation* will usually be formed in the same way as most civil engineering contracts. A tender will be submitted, which in contractual terms will normally be an offer capable of acceptance. That tender will be accepted in its original or in an amended form. Upon the unconditional acceptance (usually in writing) of the Contractor's original or amended tender, a legally binding relationship will exist.

A standard Form of Tender is provided for use in connection with these Conditions. The heading to that Form of Tender will require completion, probably at pre-tender stage.

SHORT DESCRIPTION OF INVESTIGATION:
All Site Operations, Laboratory Testing★ and Report preparation and submission★ in connection with†

....
★Delete if not required
† Complete as appropriate

This recommended Form of Tender is expressed as an offer to carry out the whole of the Investigation in accordance with the Contract Documents; it recognises that the Tender and the written acceptance of it will constitute a binding Contract.

Whether or not the tender documents prohibit making qualifications in the tender, the tendering contractor is legally at liberty to make qualifications to his tender (the employer is equally free to reject the tender or the qualifications). If qualifications have to be made, it is desirable that the Form of Tender is amended to incorporate the qualifications; in its unamended form, the accepted tender would not allow the incorporation of such qualifications.

Between the submission of the tender and the final acceptance, it is modern practice for engineers and tenderers to negotiate upon, and clarify, points of difficulty. It is essential in the interests of both employer and contractor that any points agreed upon or clarified during this period, which might alter

15

the contents of the original tender documentation, are clearly identified in a document which can be incorporated into the Contract.

As in the *ICE Fifth Edition*, the Employer can insist, by virtue of clause 9 of the *Conditions*, that a written Contract Agreement (using the Form of Agreement included in the *Conditions*) be entered into. The principal advantages of such a formality are, firstly, that it acts as a clear and enduring record of what has been agreed (of particular importance to large organisations with changing personnel) and, secondly, that it enables the parties, if they are so agreed, to place the Contract under seal (as opposed to under hand). The limitation periods (under the Limitation Acts 1939 and 1980) for the bringing of proceedings in contract are twelve years for a contract under seal, and six years otherwise (see Chapter 9).

Chapter 2
General obligations of the Contractor

There are minds so impatient of inferiority
that their gratitude is a species of revenge and
they return benefits, not because recompense is
a pleasure, but because obligation is a pain.
                              Samuel Johnson

Obligation to carry out the Investigation
The primary undertaking of the Contractor is to carry out the
Investigation and is set out in clause 8(1):

   The Contractor shall subject to the provisions of the Contract carry out
the Investigation and provide all labour materials Equipment instrumentation transport to and from and in or about the Site and everything whether
of a temporary or permanent nature required in and for the Investigation so
far as the necessity for providing the same is specified in or reasonably to be
inferred from the contract.

   This is a wide ranging obligation which is, however, subject
to the other conditions of the Contract. Whilst the Contractor
will be entitled to some contractual relief in given circumstances (unforeseeable physical conditions, clause 12; legal or
physical impossibilities, clause 13), the context of a ground
investigation Contract infers that the clause 8(1) obligation is in
practice more absolute than its civil engineering equivalent
(clause 8(1) *ICE Fifth Edition*). This context is that the object of
the Contract is to ascertain conditions which affect the ground
or properties of the ground, and that these are not known,
understood or appreciated by the parties at the inception of the
Contract. Thus, if conditions which are not so known,
understood or appreciated are encountered by the Contractor
during the course of the works, it will be difficult for the
Contractor to circumvent the wording of clause 8(1), by
reference, for instance, to clause 12.
   The Contractor is obliged to provide everything necessary
for the Investigation, whether anticipated by him at tender

17

stage or not. Clause 8(1) mirrors the common law. In one case (*Sharpe v San Paulo Railway*) 1873 LR 8 Ch App 597), the Contractor was expressly obliged to construct, for a lump sum, a railway from one terminus to another; it was necessary to excavate two million cubic yards of earthworks more than had been anticipated; the Contractor was not entitled to extra payment: his express obligation was broad enough to cover the execution both of expressly specified and of necessary work.

The extension of time clause (clause 44) does not excuse the Contractor from performance of clause 8(1); it merely extends the time in which he has to perform. Thus, if weather conditions make the Site difficult to operate upon, or result in the need for unforeseen temporary drainage, the additional costs are not normally recoverable by the Contractor.

The Contractor's obligation to provide all things required for the Investigation is limited only to such things as are expressly 'specified in or reasonably to be inferred from' any of the Contract documents (see clause 1(1)(*e*)). In general terms, the more detailed the Contract documentation is, the less is likely to be inferred; the less detailed it is, the more will have to be inferred. There are two clauses which need to be considered to appreciate the relevance and extent of this wording: clauses 11(1) and 57, which are discussed below.

Clause 11(1) states:

> The Contractor shall be deemed to have inspected and examined the Site and its surroundings and to have informed himself before submitting his tender as to the general nature of the geology (so far as is practicable and having taken into account any information in connection therewith which may have been provided by or on behalf of the Employer) the form and nature of the Site the extent and nature of the work and materials necessary for the completion of the Investigation the means of communication with and access to the Site the accommodation he may require and in general to have obtained for himself all necessary information (subject as above-mentioned) as to risks contingencies and all other circumstances influencing or affecting his tender.

Since the Contractor is deemed to have acquainted himself with these matters, it can only have been in the Contractor's, and the Employer's, interests to have done at tender stage that which he is deemed to have done upon the acceptance of the tender. In any given contract, accordingly, work, materials, equipment, instrumentation and transport which, although necessary, are not expressly specified, will be inferred in the context of what was ascertainable or foreseeable at tender stage.

The clause will now be considered in its separate components:

The Contractor shall be deemed to have inspected and examined the Site. . .

This is a positive and precise decision, 'inspection' and 'examination' are terms which direct the Contractor to take cognisance of all that which is visible on the Site and indicative of its nature and composition. They imply more than a simple site visit. The sequence of inspection and examination would vary according to the size and location of the Site. Considering widely differing sites, the recommended sequence of inspection and examination applicable to typical sites is given in Table 1 and is the subject of BS 5930: 1981, Appendix C. Unless there are any physical or other reasons prohibiting access, the Site generally should be examined on foot. The inspection should preferably be made during or immediately following adverse weather conditions, excluding those which mask or blanket the surface, such as flood or snowfall. However, examination of a mud slide would not be made under or following weather conditions likely to activate it, or of cliffs during periods of high wind force, or of refuse tips when internal combustion is active. Where it is impractical to inspect the Site on foot or the Site is extensive, aerial reconnaissance supported by stereo photography, interpreted by a trained operator, will give a fairly reliable indication of the general nature and pattern of the geology. At the time of writing, it is not possible or economic to interpret small sites by means of satellite. Even so, it must be understood that currently this type of evaluation is limited to the surface reaction of geological drift and may not reveal the material and conditions at depth.

. . . inspected and examined the Site *and its surroundings.*

The importance of this, in a general legal context, is emphasised by the Court of Appeal decision in Batty v

Table 1. Inspection of Site before tendering (clause 11(1))

In open country
(*a*) Published information to be taken to the Site should include:
written permission for access;
Site plan as provided with the Contract documents;
large scale district maps;
geological maps and memoirs;
pedological map, 'Soil Survey of Great Britain';
aerial photographs.

(*b*) Equipment should include:
camera with wide angle and telescopic or zoom lenses and supply of
high speed colour print films;
binoculars;
Abney level or similar;
ranging rod;
pocket tape (3.5 metre);
pocket compass (prismatic or equivalent);
A4 note and sketch pad.

(*c*) Observe and record:
any variation in position of farm buildings, etc., roads, tracks, gates
and boundaries from that given on the Site plan;
condition of approach roads, tracks and bridges with suitability of use
by Equipment;
suitable sites for offices, living accommodation, laboratory, etc.
together with water and foul drainage facilities;
condition of banks of rivers, streams and other water courses. Note
direction and rate of flow, signs of flood level and suitability for water
supply and disposal in drilling operations. Consider effect of possible
pollution where water supply intake and fishing rights are concerned;
vegetation with particular regard to marsh growth, indication of
seasonal ponding or high groundwater level;
signs or indications of tipping or made ground, albeit of same age;
ownership, tenancies, etc. relating to the Site plan. Verify permission
for boring, trial pits, etc. Where farming is concerned, obtain owner's
or tenant's intentions regarding crop/grazing programme, thus to
reduce compensation claims;
staff living accommodation.

Extended Sites in built-up areas
(*a*) Published information to be taken to the site should include:
written permission for access;
Site plan as provided with the Contract documents;
large scale district maps;
geological 'drift' map and memoirs;
address of local government offices and names of officers concerned.

(*b*)   Equipment should include:
camera with wide angle and telescopic or zoom lenses and supply of high speed colour print films;
pocket tape (3.5 metres);
torch for use in cellars and roof spaces, manholes, etc.;
pocket compass (to check drain, etc. alignment);
A4 note and sketch pad.

(*c*)   Before proceeding to site, obtain detailed information and history of the Site from local government offices. This should include:
copies of Ordnance Survey maps covering the recorded period of development;
details and alignment of foul and surface water drainage, power, gas and water services, including sealed drains and disused services;
Site history in terms of commercial and industrial development, waste disposal and tips;
limitations on working hours;
availability of water and electricity supply and charges;
facilities and limitations for water disposal;
if Site was subjected to bombing (1939–1945);
result of any previous investigations and details of foundation calculations and requirements for existing structures;
advice on staff accommodation.

(*d*)   Observe and record:
general and detailed state of existing structures and any variations from the Site plan;
alignment of drains and services;
limitations on Site access for equipment;
availability for Site parking, offices, stores and laboratory, including services;
Check (*c*) above.

Metropolitan Property Realisations 1978 QB 554. It is seldom that a Site can be considered as being isolated from its surroundings in terms of its physical properties. Clause 11(1) does not use the term 'adjacent' or 'immediate' surroundings. In this sense, a Site in an area subject to landslide or ancient slipped mass has potential hazard during ground investigation and inherent instability which may be activated at some time in the future to the detriment of the development. Where land-slip areas are concerned, it is possible that wash-boring may lubricate an existing slip plane to the extent that further movement takes place. Similarly, on a small enclosed site vibration from a cable percussion rig or loss of restraint by excavation of trial pits could cause damage or collapse of structures.

Therefore, the requirement for inspection and examination extends to surrounding areas in or from which conditions may exist or emanate which could affect the Contractor's performance of the Contract.

> . . . and to have informed himself before submitting his tender as to the general nature of the geology (so far as is practicable and having taken into account any information in connection therewith which may have been provided by or on behalf of the Employer). . . .

The requirement for the Contractor to have informed himself 'as to the general nature of the geology' (as opposed to 'the general nature of the ground and sub-soil' as given in the *ICE Fifth Edition*) is in keeping with the philosophy necessitating these *Conditions*. (The Employer's responsibilities for this in different types of investigation are discussed in Chapter 3; it is sufficient to say that the Contractor should inform himself, as far as is practicable, and should bear in mind the information provided by the Employer.) Variants of the amount of information which may be given to the Contractor are discussed in Chapter 1, p.8. The minimum which the Contractor is deemed to have done in connection with the geology is to have studied the Geological Survey maps (1:63360 to 1:50000). In a large investigation the experienced Contractor will verify the information provided in the Specification (though he may be required to confirm it as a part of his contractual performance). On a smaller site, or where the Contractor does not have the expertise to verify the geological information provided by the Employer, the information may be accepted by the Contractor and not verified. It is emphasised that, in this case, the non-verification applies only to 'the general nature of the geology' as revealed by study of the geological and pedological maps and memoirs, in conjunction with the site inspection.

> . . . the form and nature of the Site. . ..

This covers all the topographical, natural and man-made features which will affect the Contractor's performance.

> . . . the extent and nature of the work and materials necessary for the completion of the Investigation. . ..

Equipment is not specifically included but may be taken to be inferred. The Contractor is assumed to know and understand what it is that he is being employed to do.

> . . . to have informed himself . . . of the work and materials necessary for the . . . means of communication with and access to the Site,

and following with

> . . . the accommodation he may require and in general to have *obtained for himself* all necessary information (subject as above-mentioned) as to risks contingencies and all other circumstances influencing or affecting his tender.

This would seem to place the responsibility for all risks on the Contractor and entirely rule out the possibility of claim. However, if it were possible to foresee all the consequences of the nature of the Site the ground investigation would be unnecessary. It is not possible to foresee all the individual or localised difficulties which may be encountered. The 'general nature of the geology' and the 'form and nature' of the Site can only indicate inherent major difficulties, and it is these difficulties the Contractor is expected 'to have informed himself' of and upon which he is unlikely to have a basis for claim.

Clause 57 is set out below:

> Except where any statement or general or detailed description of the work in the Bill of Quantities expressly shows to the contrary Bills of Quantities shall be deemed to have been prepared and measurements shall be made according to the procedure set forth in the 'Civil Engineering Standard Method of Measurement' issued by the Institution of Civil Engineers in 1976 or such later or amended edition thereof as may be stated in the Appendix to the Form of Tender to have been adopted in its preparation notwithstanding any general or local custom.

The *Civil Engineering Standard Method of Measurement* (CESMM) should be examined to determine whether or not items of work or materials, which are not expressly required, are required by inference. For instance, Note 6 of Class B confirms that separate items are not required (in the Bills of Quantities) for disposal of excavated materials; thus, the Contractor would be obliged to dispose of such material although there was no expressed obligation that he should do so. However, the reverse is not necessarily correct; if an item

23

normally required by the CESMM is excluded from the Bills, it does not automatically mean that the Contractor is not obliged to perform such an item, where it is expressly specified in, or reasonably to be inferred from, another contract document.

The meaning and extent of the Investigation

Whilst the ultimate object of the conditions of contract is the submission of a comprehensive and comprehensible Report, the term 'Investigation' also includes by definition (clause 1(1)(*o*)) all work to be carried out at the Site (Site Operations) and all specified or necessary Laboratory Testing (see also Figure 1, p.7). Generally, the term 'Investigation' has been substituted in the *Conditions* for the expression 'the Works' which appears in the *ICE Fifth Edition*.

Site Operations are separately defined in clause 1(1)(*k*) to include Ancillary Works, which are also defined in clause 1(1)(*l*). Ancillary Works can be equated to the term 'Permanent Works' in the *ICI Fifth Edition* in that they are the work, materials, appliances or devices which are to remain on the Site for the Employer upon completion of the Site Operations. The Contractor is wholly responsible for the care of all Site Operations from their commencement until 14 days after substantial completion, subject to certain exceptions (clause 20(1)). (See Chapter 5, p.64). Upon completion of this stage, the Contractor must clear the Site of all equipment, surplus material and rubbish, and leave the Site and the Ancillary Works clean and in a workmanlike condition (clause 33).

When the Contractor is responsible for the execution of Laboratory Testing, whether within his own laboratory or by sub-contract, clause 8(3), a new clause, requires that approval of either is given by the Engineer in writing:

> The Contractor shall submit to the Engineer the name and address of the laboratories undertaking Laboratory Testing in accordance with the Contract and shall obtain the Engineer's approval in writing before the services and equipment of such laboratories may be used in the execution of the Contract.

It is essential that the Engineer inspects or has inspected all proposed laboratories in detail, preferably under operational conditions and by prior arrangement with the Contractor and sub-contractor. (Clause 37 gives the Engineer the power to inspect laboratories.) In many cases, he should do so in the company of the Employer and specialist advisers. Laboratory Testing is a vital part of the Investigation and the Employer should be made aware of the work, expertise and expense required to produce a single test result. This will be educational, reassuring, and may deter the Employer from demanding excessive tests. This also applies to any Engineer who is not well acquainted with laboratory work.

If the Laboratory Testing can be sub-contracted, it is advisable, before making the inspection, that the potential sub-contractor is informed of the following, within the allowable limits of confidentiality:

(*a*)   object of the Investigation;
(*b*)   type and quantity of materials likely to be tested;
(*c*)   type and number of tests likely to be required;
(*d*)   anticipated rate of delivery of samples;
(*e*)   whether samples are likely to be contaminated and type of contaminant;
(*f*)   storage period;
(*g*)   approximate period over which testing is likely to continue;
(*h*)   main Contractor's completion date;
(*i*)   that the laboratory will be inspected by the Engineer during the progress of the work.

General guidance on laboratory requirements is given by BS 5930: 1981 Section 6 and the British Standard *Code of Practice for the Identification and Investigation of Contaminated Land*, currently in draft form. The overriding consideration, however, is that the samples should be handled with care and that the cleanliness of the laboratories should be of a standard approaching that required of laboratories dealing with clinical specimens. The operators' and technicians' clothing is intended to avoid contamination of the samples and specimens which they handle as well as for their personal protection. Here the term 'sample' means that portion, or whole, removed from the ground or material and bearing a recorded sample number. A 'specimen' is that portion, of a particular size, shape or quantity, which is removed from the sample for testing. The specimen container and test sheet should bear the sample number and the suffix or sub-number of the specimen.

When nominating a laboratory for approval and when inspecting and approving it, the Contractor and Engineer should consider whether the particular laboratory can perform on time and to the necessary standards.

The check list below may be useful for this exercise. It can be extended to accommodate the object of the particular Investigation and any specialist analyses required. When toxic or radioactive substances are involved, or where public health (e.g. groundwater) is concerned specialists will prepare their own check lists.

(*a*) General conditions and cleanliness of the laboratory building, its surrounds and access.

(*b*) Heating, lighting, air-conditioning and dust removal. Floors non-slip and washable? Toilet facilities.

(*c*) Reception, handling, recording and storage of samples, including controlled temperature when required.

(*d*) Main extrusion apparatus.

(*e*) Facilities and staff responsible for visual examination of samples, with method of recording sample description.

(*f*) Specimen preparation facilities.

(*g*) Type, range, quality, calibration certificates and cleanliness of testing apparatus, together with methods of data recording. Check with BS 5930: 1981, Tables 4 and 5, BS 1377: 1975 and similar tables in the British Standard Code of Practice for the *Identification and Investigation of Contaminated Land* (currently in draft form).

(*h*) Training, qualification and experience of technicians.

(*i*) Name of person exercising control of the laboratory. What are his/her qualifications and experience?

(*j*) Check laboratory record books.

(*k*) Determine limitations of storage time for samples.

(*l*) Check arrangements for disposal of surplus and used material, particularly if contaminants are involved.

(*m*) Check that test data or report archives/microfiche exist.

(*n*) Where testing of large bulk samples will be involved, there should be covered space and suitable flooring for 'quartering', riffling, etc.

(*o*) Curing tanks for concrete test specimens.

(*p*) Examine all test pro formas likely to be required. Obtain blank copies.

The Engineer or Engineer's Representative should use the check list when they visit the laboratories to inspect testing in progress. On those occasions, he may be accompanied by the main Contractor or his representative.

Design obligations

A ground investigation is not 'designed' as such; it is planned. The Investigation as defined, however, includes aspects of design which are catered for by the *Conditions*: design of Ancillary Works, inherent (un)suitability of specified operations, design of temporary works and design or inherent (un)suitability of equipment provided by the Contractor. These elements will now be considered in that order.

Clause 8(4) and its Provisional Sum and Prime Cost Item corollary, clause 58(3), state respectively:

> Except as may be expressly provided in the Contract the Contractor shall not be responsible for the design or specification of any Ancillary Works required to be installed or constructed in accordance with the Contract.

> If in connection with any Provisional Sum or Prime Cost Item the services to be provided include any matter of design or specification of any part of the Ancillary Works or of any equipment or plant to be incorporated therein such requirement shall be expressly stated in the Contract and shall be included in any Nominated Sub-Contract. The obligation of the Contractor in respect thereof shall be only that which has been expressly stated in accordance with this sub-clause.

In the absence of express provision, the Contractor has no design responsibility for the Ancillary Works. If any damage occurs to the Site Operations as a result of a fault or omission in the design (other than one for which the Contractor is expressly responsible), the risk and responsibility for the damage rest exclusively with the Employer (clauses 20(3) and 20(4)).

Where the Contractor is required by the Contract to carry out particular operations by a specified method or using specified equipment, the Contractor will not be responsible if that which is specified is inherently unsuitable. For example, a contract may require the Contractor to use a light cable percussion rig for boring. Suppose limestone is encountered

27

which renders such a technique unsuitable and which requires rotary drilling. The Contractor is not then liable for the effects of using the light rig in limestone, at least up until the time when the presence of limestone is appreciated, when the Contractor would be entitled to a variation instruction. Where it is impossible, or where it would be illegal, for the Contractor to carry out the specified work by any method, the Contractor would be entitled to an instruction under clause 13(1) and/or clause 51(1).

By clause 8(2), the Contractor takes 'full responsibility for the adequacy stability and safety of all Site Operations Laboratory Testing and methods of working.'

This can be equated with the Contractor's responsibility for Temporary Works under the *ICE Fifth Edition*. As discussed above, this responsibility does not extend to the design of the Ancillary Works.

Whilst the Contractor is required to submit method statements to the Engineer for approval (see clause 14(1)), the Engineer's approval does not satisfy or relieve the Contractor of his responsibility under clause 8(2).

The sub-clause clearly imposes on the Contractor a duty to carry out all operations at the Site in a way which is adequate (for the purposes of the Investigation), stable (so that the Site or other property is not rendered unstable) and safe (for operatives and visitors as well as other people and property whose safety might be directly affected by the Site Operations).

The Contractor effectively warrants the suitability of the Laboratory Testing. This expression is defined (clause 1(1)(*m*)) as 'the testing operations and processes necessary for the preparation of the Report'. The combination of clauses 1(1)(*m*) and 8(2) adds up to a performance requirement whereby the Contractor undertakes that the testing operations and processes which he effects will not only be fit for the purpose but also can, and will, do that which the Specification, Schedules and Drawings require.

So far as responsibility for Equipment is concerned, the Contractor's obligation is set out in clause 36(3):

All equipment shall be of the respective type and kind suitable for the proper execution of the Investigation in accordance with the Engineer's instructions and shall be calibrated and subjected to such calibration tests as may be necessary to ensure performance in accordance with the requirements of the Contract.

Equipment is defined in clause 1(1)(v):

'Equipment' means any appliances or things of whatsoever nature required temporarily for carrying out the Site Operations but does not include anything which forms part of the Ancillary Works.

Clause 36(3) can be considered in its two aspects: firstly where Equipment is not expressly, and secondly where it is expressly, specified in the Contract documents. In the former case, the Contractor must provide Equipment which is in all respects suitable for the purposes of, and for the achievement of the desired result of the Investigation. In the latter case, the Contractor is obliged to provide the specified Equipment even if it is inherently unsuitable; however, subject to that, the Equipment must be capable of performing to the limits of its manufacturer's recommendations, and it must be in good working order. If the Equipment is capable of acceptable modification to render it inherently suitable, the Contractor would be obliged to procure such modification.

A grey area exists in law as to whether or not a Contractor or an Engineer/Employer owes any duty to the other or to third parties if he discovers that the other's methods of working or design are deficient or unsafe. Contractually, the *Conditions* do not impose any express duty on either party to the Contract to inform each other, although as soon as the Engineer knows of such events he has wide powers (clause 51(1), variations; clause 13(1), general instructions) to resolve any discovered problem.

Either party will owe to third parties a duty of care (in tort and independent of contract) to carry out their respective functions with reasonable care and skill. For example: the Contractor is expressly required to deposit all excavated material at the edge of the site adjacent to a school; he discovers what he ascertains is acid-contaminated material, which he deposits at the specified point; rain causes acid to flow into the school area so that children are injured; the Contractor is liable to the children for the personal injuries.

A code is laid down by clause 22 to regulate the apportionment of financial responsibility for most such events between

Contractor and Employer. The knowledge which Contractor or Employer/Engineer should have had as to events which could foreseeably cause injury or damage is an important factor in such determination.

Setting-out

The obligation to set out follows the procedures in the *ICE Fifth Edition*, with one important innovation. The Contractor, who is responsible for the accurate setting out of the Site Operations, must 'use the bench marks and on-site reference points established by the Engineer pursuant to clause 7(1) to determine and record levels and set out' (clause 17). By clause 7(1), the Engineer must designate or establish appropriate bench marks and on-site reference points to enable the Contractor to set out. If the Contractor sets out incorrectly in relation to the information provided by the Engineer, he is responsible. If, however, errors arise as a result of incorrect or inaccurate information provided by the Engineer, the costs of, and occasioned by, correction are borne by the Employer.

Responsibility for sub-contractors

Sub-contracting does not relieve the Contractor of any obligation or liability which is imposed upon the Contractor by the Contract (Clause 4). All sub-contracting (save for labour-only piece-work) must be consented to by the Engineer before it occurs. Any laboratory which the Contractor proposes to utilise must be approved by the Engineer before it starts work (clause 8(3)). Delays occasioned by sub-contractors will not attract extensions under Clause 44, unless there are extraordinary surrounding circumstances. The Contractor must procure for the Engineer facilities for access at any time to the laboratories, workshops and places of manufacture or storage of any sub-contractor, nominated or otherwise (clause 37). The Contractor is obliged to ensure that any sub-contract for the execution of any part of the Site Operations contains the property-vesting sub-clauses of clause 53.

The Contract makes express provisions for Nominated Subcontractors which are not materially different from those in the *ICE Fifth Edition*.

Programme and progress
Clause 14(1) states:

Within 21 days after the acceptance of his Tender the Contractor shall submit to the Engineer for his approval a programme showing the order in which he proposes to carry out the Investigation and thereafter shall furnish such further details and information as the Engineer may reasonably require in regard thereto. The Contractor shall at the same time also provide in writing for the information of the Engineer a description of the arrangements and methods which the Contractor proposes to adopt for the carrying out of the Investigation. The programme submitted by the Contractor pursuant to this sub-clause shall take into account the period or periods for completion of the Investigation or different Sections thereof and the periods required by the Engineer for approval of testing schedules and reports which are provided for in the Appendix to the Form of Tender.

Whilst the period of 21 days can be changed by prior arrangement, most Contractors will find it practicable to submit the programme envisaged by clause 14 on time. Contrary to common belief, this programme needs to show only the proposed order (or sequence) of the period, or periods, of completion for the various stages of the Investigation; it does not need to show calendar dates. The programme must also show the period for approvals of testing schedules and reports which should have been inserted in the Appendix to the Tender. However, the Engineer is entitled to require the Contractor to submit details of the precise dates envisaged by the Contractor. It is important for both Employer and Contractor that the programming proposals are realistic and achievable. There are three main respects in which the programme has contractual relevance:

(a) The granting by the Employer of possession of portions of the Site is related expressly to the programme. Possession is required to be given from time to time only in such portions as will enable the Contractor to proceed in accordance with the programme (clause 42(1)).
(b) The programme can be used legitimately by both Contractor and Engineer as a relevant yardstick against which to assess extensions of time under clause 44 and to monitor progress (clause 14(2)).
(c) The timing of the release of further information and

31

instructions by the Engineer (clause 7(1)) can be regulated largely by reference to the programme. Generally, the Contractor's right to the timely provision of information can justifiably relate to the programme.

The Date for Commencement of the Investigation is that notified by the Engineer in writing (clause 41). The date must be within a reasonable time after the creation of the Contract (acceptance of Tender). In most cases, the date can reasonably be notified to be after the submission of the clause 14 programme and method statement by the Contractor. Once commenced, the Contractor must proceed at a sufficient pace to ensure that substantial completion is achieved within the time(s) for completion specified in the Appendix to the Tender (clause 43), subject to the provisions for the extension of that time (clause 44). The context of a ground investigation Contract allows the use of sectional completion, as the general rule, since the conclusion of at least some of the Site Operations will precede by some time the submission of the Report.

The Contractor is entitled to extensions of time under clause 44, where he has been delayed beyond the time(s) for completion of the whole (or sections) of the Investigation, by four classes of occurrence.

(*a*) Variations or increased quantities.
(*b*) Specific causes expressly referred to elsewhere in the *Conditions* (clause 7(4): late release of information, failure to establish bench-marks and on-site reference points; clause 12(3): unforeseeable physical conditions and artificial obstructions; clause 13(3): Engineer's instructions; clause 14(4): late consent to method statement or unforeseeable instructed change in sequence; clause 14(5): delayed approval of testing schedule or reports; clause 27(6): street works variations; clause 31: facilities for Employer's other contractors; clause 40: suspension; clause 42(1): delayed possession; clause 59B(4): forfeiture of nominated sub-contract).
(*c*) Exceptional adverse weather conditions. Weather conditions are adverse if they tend to delay or disrupt the Contractor's progress. However, such conditions are not exceptional merely because they delay or disrupt the

Contractor. Exceptional means exceptional in severity or longevity relative to the time of year.

(*d*) Special circumstances. A circumstance is not special simply because it delays the Contractor. A special circumstance is (it is considered) a circumstance beyond the reasonable control of the Contractor which was not foreseen or reasonably foreseeable prior to the Contract.

Reinstatement of boreholes, trial pits, shafts and adits

The reinstatement of boreholes, trial pits, shafts and adits to a condition and by methods which will avoid injury to persons or damage to property is an important part of satisfactory execution of the Contract. It is mentioned in several clauses which will now be discussed. It will probably be the case in most Contracts that, even if the need for reinstatement is not expressed, it will be reasonably inferrable from the Contract documentation. Thus the Contractor would be required to reinstate by clause 8(1). Indeed, the CESMM (Class B, Note 6) does not require separate identification in the Bills of Quantity for reinstatement of surfaces. However, there is no clause which requires the Contractor specifically to reinstate to the satisfaction of the Engineer. To avoid confusion, the requirement, together with details of the method of reinstatement should therefore be included in the Specification and thus become an express requirement within the Contract. It is essential that special measures for reinstatement are expressly specified in the Contract, otherwise the Contractor will not be obliged to put such measures into effect without a variation order.

Methods generally applicable to reinstatement are given by BS 5930: 1981, clause 18.9 'Backfilling excavations and boreholes' and supports the necessity for detailed attention to this operation.

It is not possible to detail here the specific requirements for each situation, since these are infinitely variable and should be specific for the Site concerned. For example, the excavation of a trial pit in arable land may well require that topsoil is set aside in a stockpile, or stockpiles, in the natural sequence of pedological nutrient value so that there is no reduction in cropping intensity. Similarly, the backfilled soil should be in reverse sequence to that excavated, so that the indigenous strata

33

pattern is not interrupted and the compacted density of the backfilled soil equates with the indigenous density, thus avoiding excessive consolidation. Where excavations in highways and other roads, tracks or bridleways are concerned, it will be necessary to meet the requirements of the highway authority and in safety, where statutory or bye-law requirements do not apply, to slab over the backfilled material, thus ensuring that consolidation is not transmitted to the surface.

It is possible that shafts and adits made during the Investigation may be designated for future use in construction or the installation of Ancillary Works, in which case temporary support will be specified. Should this not be so, it will be necessary to backfill in a manner that will ensure that subsequent collapse or sub-surface water tankage will not occur and it is probable that pumped concrete will need to be specified.

It is also important to avoid pollution of aquifers or natural water bearing zones used for water supply from wells, spring transfer or pumping. In such cases, reinstatement of investigory excavations by concreting fully should be considered in the Specification.

The effect of clauses 8(2) and 18 can be to require the Contractor to reinstate where it is necessary for purposes of stability and safety.

Clause 13(7) allows the Contractor to backfill, without instruction, boreholes which he has found it necessary to continue or alter on his own initiative, under Clause 13(4). There is a detailed consideration of Clauses 13(4) to 13(7) in Chapter 7.

The Contractor is obliged, if so directed, to open up at his own expense, any trial pits which have previously been reinstated and which were not previously offered to the Engineer for inspection and/or which did not comply with the Contract. The Contractor would then be obliged further to reinstate at his own expense (clause 38(2)).

A Certificate of Completion of a section or part of the Investigation is deemed, unless it expressly states otherwise, to exclude ground or surface reinstatement (clause 48(4)). This sub-clause recognises that reinstatement may not be fully effected at the earlier stages of completion.

By clause 49(2), the Contractor is required to carry out further works of reinstatement where the need arises from 'defects imperfections shrinkages damage subsidence of backfill or other faults resulting from execution of the Site Operation, occurring by the end of the Period of Maintenance.

Special provisions are made for the temporary reinstatement of highways which have been broken into for the purposes of the Site Investigation (clause 49(5)).

(The Contractor's responsibilities in connection with quality are considered in Chapter 4; those relating to care, security and superintendence in Chapter 5; and those to do with completion, maintenance and reports in Chapter 6.)

Chapter 3
General obligations of the Employer and the Engineer

For 'tis the sport to have the engineer
Hoist with his own petard: and it shall go hard
But I will delve one yard below their mines
And blow them to the moon.
William Shakespeare

The Employer's responsibility for the Engineer

It is crucial that a good working relationship is established at the earliest stage (even at tender stage, if possible), because site investigation is a complex art, and because the Employer and the Engineer will be working together closely throughout the Contract.

There is a tendency for some Engineers to interfere with the Contractor's performance to such a degree that expense and delay are caused by continuous over-attention to non-essential details. It should be understood that the object of the Investigation can be achieved in several ways, most of them known to an experienced Contractor. This does not mean that the Engineer's supervision should be lax. However, the Engineer should not 'ride his back' unless it is obvious that the Contractor is inexperienced. The Contractor, in turn, should not be tempted to make an 'easy penny'.

Invariably, the Engineer will be engaged (in the case of a private firm) and employed (in the case of an employee) by contract to the Employer. By that contract, the Engineer will owe to the Employer a duty to exercise reasonable care and skill; his duty is often expressed (e.g. clause 5.1), *Association of Consultant Engineers (A.C.E.) Conditions of Engagement.* (Agreement 1 for Report and Advisory Work) and, where not expressed, will always be implied. The Engineer (where independent) will be vicariously responsible for the acts and omissions of his employees acting in the course of their employment (for instance, as Engineer's Representatives). The Employer could be liable vicariously for the acts and omissions of his employee-Engineer acting in the course of his employment.

The Engineer is certainly, in the above sense, the 'Employer's man' (or woman). The acts, or the failures to act, of the Engineer will directly affect the Employer. By appointing a person or firm as the Engineer under the *Conditions*, the Employer confers upon that person, or firm, legal authority to act for, and on behalf of, the Employer and thus to bind the Employer. Accordingly, if the Engineer orders a variation under clause 51(1) which would attract additional payment, the Employer is liable to pay such a sum to the Contractor, even if the variation was not necessary, or arose as a result of some failure on the part of the Engineer. The Employer may have some recourse in law against the Engineer, but that factor does not affect the Employer's obligations to the Contractor.

If there are to be any restrictions imposed upon the functions of the Engineer under the Contract, this can only be achieved by express agreement between the Employer and the Contractor. In the absence of such an agreement, the Contractor is entitled to proceed upon the basis that the Engineer does have the authority allowed in the Contract. Indeed, if the Employer interferes with, or seeks unilaterally to restrict, the functions of the Engineer after the Contract has been made, he will be in breach of contract.

Although the Engineer is the employee or agent of the Employer, he has an independent function which is acknowledged both by tradition and in the *Conditions*. This function has been termed 'quasi-judicial' or 'quasi-arbitral'. It arises where the Contract requires the Engineer to form an opinion (e.g. clause 48(1), opinion as to whether the Investigation is substantially complete), to make an assessment (e.g. clause 44, extensions of time), or to certify (e.g. clause 60(2), interim payments). As agent or employee, the Engineer is obliged to form the opinion, to make the assessment or to certify. However, the content of the opinion, assessment or certificate is the subject matter of the independent function. Some Engineers mistakenly believe that this attribute entitles them to do what they believe is fair and reasonable without reference to the Contract: that is incorrect; they must apply the particular contractual context in which the function arises. The most important aspect of the Engineer's quasi-judicial position is

37

honesty and impartiality. Before forming his opinion, making the assessment or certifying, the Engineer is entitled to invite representations from both sides. In this quasi-judicial capacity the Engineer is not entitled simply to do what he is directed to do by the Employer. The employee-Engineer is undoubtedly in a difficult position in these circumstances; in the absence of contrary express terms of the contract of employment, the Employer must give his employee the right to function independently as Engineer.

There is nothing which the Engineer does, either as employee or agent or as independent functionary, which is final and binding. Any decision, opinion, instruction, direction, certificate or valuation of the Engineer is subject to review and revision in arbitration (see Chapter 9).

There are many instances in the Contract where the acts or omissions of the Engineer legally (and financially) affect the Employer. For example, the failure of the Engineer to establish bench marks either promptly or at all may have a direct consequence (an extension of time and an increase in the Contractor's costs and overheads, clause 7(4)). If the Engineer takes longer to approve the testing schedules than the times given by the Appendix to the Form of Tender, the Contractor may be entitled to an extension of time (clause 14(5)).

It is an implied term of the Contract that the Employer himself, his servants or his agents (including his Engineer) should act so as not to hinder or prevent the Contractor from complying with the Contract. Whilst there are some clauses which provide for extension and reimbursement for certain more obvious acts of hindrance and prevention (e.g. late instruction by Engineer, clause 7(3)), there are other matters not so covered which will result in a claim for damages by the Contractor. For instance, where free-issue materials or apparatus are to be provided by the Employer, and what is provided is deficient or is delivered late, the Employer would be in breach of the implied obligation. This is considered later.

The honesty, impartiality and efficiency of the Engineer and the provision of reasonable co-operation by the Employer are crucial to the proper administration of the Contract.

Responsibility for the design

Since a ground investigation is not generally designed as such (it is planned or specified), meaning has to be given to the term 'design of the Investigation' which appears in clause 20(4). The term, specifically in that context and generally in the Contract, means and includes the design of the Ancillary Works and the mandatory specification of particular work, apparatus, devices and materials which the Contractor is obliged to use.

The Employer is not responsible *vis-à-vis* the Contractor for that part of the Investigation which is designed by the Contractor (see Chapter 2, p.27). The Employer may have liabilities at common law to third parties who are harmed by the activities of the Contractor; provision is made for indemnities and insurance against such a risk (clauses 21 to 24).

It is clearly established that the Employer does not warrant that the Investigation can be carried out in accordance with the Contract documents (Thorn v London Corporation 1876 1 App Cas 120, Tharsis Sulphur and Copper Co v McElroy 1878 3 App Cas 1040, Pearce v Hereford Corporation 1968 66 LGR 647). Clearly the Contractor must carry out the Investigation no matter how impracticable or difficult it happens to be. The effect of this legal principle has been mitigated to a large extent by the *Conditions*, in particular by clauses 13, 20 and 51.

The Contractor's obligation is to 'carry out the Investigation in strict accordance with the Contract', 'Save insofar as it is legally or physically impossible' (clause 13(1)). This clause will be considered in detail in Chapter 7. If the Employer's or Engineer's design of the work is such that, assuming that the Contractor fully complies with the Contract, it cannot physically or legally be effected by the Contractor, the Engineer must issue instructions to overcome the impossibility. This obligation is not expressed as mandatory (save under clause 51(1) where the solution to the impossibility 'is a variation), and arises from the implied term of non–hindrance and non–prevention (see p.39). Physical impossibility may arise when borehole or trial pit positions and depths and types of sampling have been specified, and some aspect of topography, geology, seasonal groundwater or sea condition renders the specified sampling impossible. Instructions would have to be given to overcome this.

It is laid down in the Contract that where damage, loss or injury occurs to the Site Operations between commencement and 14 days after substantial completion, as a result of fault, error or omission in the design of the Investigation, the risk and the responsibility for paying for the damage, loss or injury rests with the Employer (clause 20(4)). This is considered in detail in Chapter 5. The risk, however, is not so great as under the *ICE Fifth Edition*, since less design is involved, and there are fewer permanent works.

If it is discovered that the Drawings, Specification or instructions of the Engineer will bring about, or have caused, a non–compliance or a breach of statute, statutory regulations or bye–law, the Engineer must issue instructions to prevent or rectify this (clause 26(2)(*b*)). For example, the innocent Contractor may be required by the Contract to drill a borehole where it is bound to encounter a public utility service (e.g. cable or pipe). The Engineer must issue an instruction to avoid this (if he discovers it in time) or to rectify the effects (if it has happened), since there will probably have been a contravention of statute.

Where design changes are necessary to any part of the Investigation, the Engineer is bound to issue the appropriate variation order on behalf of the employer, as discussed on p.37. The opening terms of clause 51(1) are mandatory: 'The Engineer shall order any variation . . . that may in his opinion be necessary for the completion of the Investigation. . . .'

Since the word 'shall' is used, since the opinion of the Engineer is subject to review in arbitration, and since the following wording in clause 51(1) is discretionary for 'desirable' variations, necessary design changes must be ordered by the Engineer even if the work could proceed unvaried.

Establishment of bench marks and reference points
Clause 7(1) states:

> The Engineer shall designate or establish bench marks and on–site reference points to enable the Contractor to set out the Site Operations in accordance with Clause 17.

The execution of this requirement has far reaching effects, not only on the accuracy of the ground Investigation, but on any construction which may follow. The Engineer may consider that in order to 'designate' a bench mark, he need only refer to an Ordnance Survey plan, select the bench mark most convenient to the Site, state that this bench mark shall be used, and indicate it on the Site plan included with the Contract documents. This may satisfy a point of law, but it would be unwise, especially where large projects are concerned, to take this responsibility lightly. The Engineer should verify the position, and the most recently calibrated level of the bench mark, with the Ordnance Survey, and let the Contractor know, in writing, that this has been done. He should also check, that the bench mark is in the position shown on the map and that it is convenient to the Site in terms of the distance, ease and accuracy with which the true level can be transferred to a workable position on the Site. On an extended Site, such as a new road alignment, it may be necessary to check several bench marks convenient to the proposed alignment.

This true level is essential to the accuracy of the ground Investigation Report and to the subsequent works which will be based upon the information which the Report provides, particularly where groundworks are concerned. In such circumstances, the Engineer may wish to establish a permanent Site bench mark which will serve both the Investigation and the construction which follows. He will thus ensure that there will be no errors related to the level data below ground in the construction, or in the specification for the construction. Although clause 7(1) states that the Engineer may either designate or establish, he must do that which is appropriate to the particular Investigation.

The requirements for setting-out are given by clause 17 and obviously follow on from clause 7(1), discussed above.

The Contractor shall use the bench marks and on-site reference points established by the Engineer pursuant to Clause 7(1) to determine and record levels and set out the positions of the Site Operations. The Contractor shall be responsible for the true and proper setting-out of the Site Operations and for the correctness of the position levels dimensions and alignment of all parts of the Site Operations and for the provision of all necessary instruments appliances and labour in connection therewith. If at any time during the progress of the Investigation any error shall appear or arise in the position levels dimensions or alignment of any part of the Site Operations the Contractor on being required to do so by the Engineer shall at his own cost rectify such error to the satisfaction of the Engineer unless such error is based on incorrect or inaccurate bench-marks on-site reference points or

41

incorrect data supplied in writing by the Engineer or the Engineer's Representative in which case the cost of rectifying the same shall be borne by the Employer. The checking of any setting-out or of any line or level by the Engineer or the Engineer's Representative shall not in any way relieve the Contractor of his responsibility for the correctness thereof and the Contractor shall carefully protect and preserve all bench-marks sight rails pegs and other things used in setting out the Site Operations.

This clause is to some extent self-explanatory, but it will be noted that it refers to the bench marks or reference points as having been 'established' (not merely designated). This clause recognises that the responsibility for any errors, however occurring, in connection with the bench marks and reference points rests firmly with the Employer and may eventually be the basis of a claim by the Employer for professional negligence against the Engineer.

This clause applies to all sites, regardless of size or position, but the extent and detail to which its requirements are executed are dependent upon the area which the Investigation covers. Maintenance of accuracy by means of checking applies to all sites; monitoring and checking as a continuing obligation of both the Engineer and the Contractor are more applicable to the larger or extended site. On such Sites, it is likely that temporary on-Site reference points and setting-out, such as wooden or unprotected pegs, will probably be removed for reasons which have nothing to do with the Investigation. They may also be displaced by normal traffic to the Site or by traffic engaged in the work on the Site. The construction of semi-permanent reference points, protected against normal hazards, is therefore preferable and likely to maintain accuracy more economically than repeated replacement and re-levelling. In subsequent claim for error, the responsibility for the subsidiary reference indicated by the clause may rest initially with the Contractor, who must 'carefully protect and preserve all bench-marks'. This latter provision does not, however, refer to 'on-site reference points' established by the Engineer; therefore, the Contractor has no responsibility to protect or preserve them beyond ensuring that no damage or displacement is caused by his own operations.

Provision of information

Breakdowns in communication and ineffective communication between Engineer and Contractor are common problems. In a ground investigation contract, particularly where phased investigation and testing, or continuous monitoring are being carried out, these problems can be critical. On many contracts, the investigations will have to be regulated, modified and adjusted often, and occasionally on an hour-to-hour basis. Reliable lines of communication between Engineer and Contractor must therefore be established as soon as possible at both Site and senior level.

Generally, it will therefore be necessary for the Engineer, or the Engineer's Representative on the larger jobs, to be physically on the Site on a regular basis and, on the smaller jobs, to be available at short notice. The Engineer must in appropriate cases maintain a continuous presence on the Site, and also insist upon regular visits to laboratories.

It has been recommended in Chapter 1 (p.12) that all available information should be provided to the Contractor at tender stage. If this is not possible, there may be a great burden on the Engineer, after the acceptance of tender, in connection with the supply of information.

It will invariably be necessary for the Engineer to provide further information and instructions during the course of the Investigation. Clause 7 provides the contractual code. Sub-clause (2) states:

> The Engineer shall have full power and authority to supply and shall supply to the Contractor from time to time during the progress of the Investigation such modified or further drawings and schedules and instructions as shall in the Engineer's opinion be necessary for the purpose of the proper and adequate execution of the Investigation and the Contractor shall carry out and be bound by the same.

This clause imposes a mandatory obligation on the Employer, through his Engineer, to provide this further information. Failure to comply with this obligation will entitle the Contractor not only to the contractual relief set out in clause 7(4), but also possibly to damages.

The obligation is to supply the information 'from time to time', but at a time reasonable in all the circumstances. As to what is reasonable, four areas are particularly relevant:

43

(*a*)   The programme and other information provided by the Contractor pursuant to clause 14. Thus, for example, where a Contractor whose programme indicated extensive and time-consuming probing first, and boring of the first boreholes some months later, requests, on day 1, information regarding the precise location of each borehole, the Engineer may be neither able, nor obliged, to provide the information immediately.

(*b*)   The Contractor should have given adequate written notice of his further information requirements (clause 7(3)). Adequate notice means adequate in relation to progress on Site and programme, so as to allow the Engineer a reasonable time to provide the requested information.

(*c*)   The actual progress of the Investigation at the time at which the information is said to be required is relevant to the time at which the information is required.

(*d*)   If the Investigation is a phased investigation, in which it is clear from the Contract that the Engineer will make decisions based on interim findings and reports submitted by the Contractor, further decisions need not generally be communicated until sufficient findings and reports have been submitted for the making of such decisions.

The further 'drawings, schedules or instructions' referred to in clause 7(1) are modifications or additions to the Contract Drawings (see clause 1(1)(g)); 'Schedules' can be categorised principally as information relating to Site Operations and Laboratory Testing (clause 1(1)(b)); 'Instructions' are any of the instructions which the Engineer is empowered to issue under the *Conditions* (e.g. clause 13(1), clause 51(1)).

The requirement (clause 7(3)) that the Contractor must give adequate notice of his further information requirements is not a precondition to the Engineers' obligation to supply information to the Contractor under clause 7(2).

Approvals of testing schedules and Reports
The testing schedules may need to be in three parts for each stage of the Investigation:

(*a*)   *In situ* tests;

(*b*) laboratory tests, including chemical analysis of contaminants;

(*c*) the operation of Ancillary Works.

These may be prepared by the Engineer where the Investigation is designed and planned by him, or by the Contractor, when only the object and stated objectives have been given by the Engineer and the planning of the Site and ground Investigation is the Contractor's responsibility. The 'object' of the Investigation is to obtain or present all the information required in accordance with the conditions of contract. The 'stated objectives' will vary with the type of Investigation. They relate to the completion of the different parts, or Sections, of the Investigation. They could be, for example, the desk study, the Site work or Laboratory Testing.

In the first case, the schedules prepared by the Engineer will be discussed with the Contractor to determine their feasibility and what specialised testing equipment and methods he proposes to employ. In the second case the Contractor will propose particular tests for approval by the Engineer, who will confirm that they will provide the information he requires, and will not involve an unnecessary amount of testing.

It is not possible in either case to detail with certainty the number of, or position at which, *in situ* tests are to be made or samples extracted for laboratory tests. The type of tests, where they should be made, and when the samples should be extracted, will depend upon what changes in soil type, or changes in condition within that type, are encountered during boring or trial pit excavation. If the preliminary investigation has included probing, it will be possible to give more detail of type, depth and sequence of test. In forensic investigation to determine the reason for occurrences such as foundation failure of structures, landslip or embankment failure in roadworks, a different approach is taken: continuous, minutely detailed recording of the strata and conditions encountered, probably with continuous sampling and *in situ* testing, is necessary. This will probably be detailed by the Engineer who will be responsible for the design of the remedial works. All these requirements, are discussed in Chapter 4.

The type, sequence and required content of Reports is discussed in Chapter 6. Clauses 1(1)(*n*) and 14(5) of the

45

Conditions, and the Appendix of the Form of Tender refer only to the draft and final Report. It is unlikely that these will be the reports required, since it is advisable that most investigations should be staged in order to provide information continuously throughout the Investigation. Staged reporting makes for economy of effort and expenditure and means that the information required may have been provided before the full Investigation is completed.

Reports may have to be presented at the completion of the desk study and preliminary reconnaissance and, if the Investigation proceeds beyond this, reporting will continue with the submission of the Site (or driller's) log as illustrated by BS 5930: 1981, Figures 20 and 21, and the preliminary boring logs giving a soil description from visual examination and graphical representation of the strata encountered, together with the results of *in situ* tests. These will not include the result of laboratory soil tests since the detail given will indicate what primary or further Laboratory Testing is required. From the laboratory results a final soil classification will be produced together with descriptions, illustrated by the British Soil Classification for Engineering Purposes BSCS Classification Symbols. The observation and field testing necessary to produce the preliminary logs is given by BS 5930: 1981 Table 6 and the final classification from laboratory results by Table 8 of that code of practice.

The preliminary logs supported by an interim Report, will form part of a 'stage'. In many cases, such as an investigation for land use, no further information will be required. In such a case it may be necessary, given storage limitations, to retain samples for extended testing in order to provide design data for structures or construction which the topography, geomorphology and seasonal condition of the Site permit.

Whatever the type, content, form and timing of the Reports required, they must be specified in the Contract documents.

At each of the specified stages, the Engineer must give his approval, or reasons for disapproval, within the times agreed (e.g. in the Appendix of the Form of Tender) or, if none have been agreed, within a reasonable time. (If the Engineer fails to give approvals within agreed (or reasonable) times, extensions of time are awardable (clause 14(4)), and the Contractor should be entitled to any resulting loss as damages for breach of the implied term of non-hindrance.

Possession of the Site

The situations, geography and circumstances of sites to be investigated will vary widely. Sites may be large, small, contained or spread out; occupied or unoccupied; some may be virgin while others may have existing structures or services. Some sites may be in the course of construction, others may be completely or partly under water. Some sites may not be owned by the Employer. In their unamended state the *Conditions* impose strict obligations on the Employer as to the granting of possession. Accordingly, except in the case of an unoccupied virgin site owned by the Employer, special provisions must be included in the Specification or as clause 72, if the Employer wishes to avoid the rigours and effects of the *Conditions*.

The Site is defined in clause 1(1)(*q*):

(*q*)   'Site' means the lands and other places on under in or through which the Site Operations are to be executed and any lands places or access thereto provided by the Employer for the purposes of the Contract;

To avoid doubt, it is common, and prudent, to delineate the Site on a Contract plan. The 'Site' cannot be varied in position by variation order under clause 51(1), but it may be extended or reduced.

Clause 42(1) of the *Conditions* is the principal clause governing the granting of possession:

Save insofar as the Contract may prescribe the extent of portions of the Site of which the Contractor is to be given possession from time to time and the order in which such portions shall be made available to him and subject to any requirement in the Contract as to the order in which the Investigation shall be executed the Employer will at the Date for Commencement of the Investigation notified under Clause 41 give to the Contractor possession of so much of the Site as may be required to enable the Contractor to commence and proceed with the Investigation in accordance with the programme referred to in Clause 14 and will from time to time as the Investigation proceeds give to the Contractor possession of such further portions of the Site as may be required to enable the Contractor to proceed with the Investigation with due despatch in accordance with the said programme.

The clause acknowledges that certain special provisions may have been agreed in other Contract documents.

'Possession' normally means the exclusive and unrestricted right to be on the Site. In these conditions, sufficient possession must be given to enable the Contractor to comply, without hindrance, with all the terms of the Contract. For example, if a pasture field is the Site and if the Contractor, whilst allowed to go on to the field, is constantly impeded by the lawful grazing of the owner's or tenant farmer's cows, that is not proper possession in accordance with clause 42(1) (unless express provision was made for this elsewhere in the Contract).

Possession is to be granted, from time to time, only of such portions of the Site as are necessary to enable the Contractor to comply with his clause 14 programme. Obviously, possession will relate almost exclusively to the Site Operations. Once the Site Operations are complete, the Employer must be in a position to procure further possession or further access if such is required (for instance, for maintenance, further monitoring of Ancillary Works or, if necessary, for further investigative works.

If any part of the land comprising the Site is occupied during the Investigation by others, the Contractor must generally take care to avoid damage to their persons and their property and to the land which they occupy. However, if such damage is the unavoidable result of the Site Operations carried out in accordance with the Contract, the risk is the Employer's (see clause 22). For example, if the effect of boring operations, carried out correctly in accordance with the Contract, is to undermine or otherwise adversely affect the stability of a building on the Site occupied by a third party, the Employer will be responsible *vis-à-vis* the Contractor (see Chapter 5, p.69).

In a ground investigation contract, the Engineer or Employer will often need to bring specialists, consultants or other contractors onto the Site during the course of the Site Operation. Procedures exist to allow this (clauses 31 and 37), subject to one proviso (unless the Contract provides otherwise): to the extent that the Contractor incurs delay or cost as a result of being disrupted or prevented by the specialists, consultants or other contractors from proceeding properly, the

Employer will be liable in terms of extension of time and reimbursement (see clause 31(1) and p.39).

Difficulties with access to the Site and problems with third parties' rights of way and rights over the Site will often be encountered in ground investigation. The responsibility for ensuring that the Contractor's performance is not affected rests entirely with the Employer.

Obligation to pay

The primary commercial obligation of the Employer is to pay to the Contractor the sums due under the Contract in the amounts, and at the times, laid down in the Contract. The sums due are related primarily to the rates and prices in the Bills of Quantity which are part of the Contract. They are to be determined on a measure-and-value basis (or by 'admeasurement', clause 56(1)). Commonly, the quantities for ground investigation will be even less determinate at tender stage than those for civil engineering projects. The task of measurement is inherently more difficult on a ground investigation than on civil engineering work, in this sense: on the latter, the bulk of work is visible and thus physically measurable; on the former, much of what has to be 'measured' (apart from trial pits) is not visible. For instance, MacKintosh probes may be specified, measurable by depth rather than payable on an item basis; unless the Engineer's staff are present throughout each probe, the Engineer would have to rely almost exclusively on the accuracy of the Contractor's record. Bearing in mind this difference, it is crucial that procedures are specified or agreed at an early stage, to ensure that accurate, verifiable records are kept.

Interim certification by the Engineer is conditional upon the submission, after the end of every calendar month, by the Contractor, of details of estimated amounts which the Contractor considers he is entitled to up to the end of that month (clause 60(1)). It may well be that relatively large sums are due early in the Site Operations; for instance, often the priced Bills of Quantity will have an item 'Equipment to Site', to which the successful Contractor has attributed a large lump sum; if all the Equipment is provided at the Site early on, the whole sum will be due. The Contractor's submission can include not only the value of work and services provided, and goods and materials delivered to Site, but also the estimated amount of any claims.

Within the 28 days following the monthly submission of the Contractor, the Engineer must have certified and the Employer must have paid the amounts which the Engineer considers are due against the submission (see clause 60(2)). The Engineer is entitled to take into account deficiencies on the part of the Contractor which affect the value of the Investigation (clause 60(7)) and to make adjustments in his certification. He must obviously relate payment to the billed rates and prices.

The final certificate and payment is to be made within three months after the date of the Acceptance Certificate (clause 60(3)). This Acceptance Certificate is issued after the expiry of the Period of Maintenance and any making good, but also after Certificates of Completion have been issued by the Engineer in respect of all Laboratory Testing and all Reports called for by the Investigation.

Provision is made for retention monies (clause 60(4)), but in this Contract it is optional: the wording of the *Conditions* does not have to be altered if no retentions are required; the Appendix to the Form of Tender must have been appropriately altered as indicated in the Form of Tender:

Retention in accordance with Clause 60(4) is/is not[g] to be deducted.

[g] Delete as appropriate.

Sanctions available against the Employer for failures to certify or pay properly or on time are fourfold.

(a) Interest is expressly chargeable at: a rate per annum equivalent to 2 per cent above the Average Base Lending Rate of Lloyds Barclays National Westminster and Midland Banks current on the date upon which payment first becomes overdue. In the event of any variation in the said Average Base Lending Rate being announced whilst such payment remains overdue the interest payable to the Contractor for the period that such payment remains overdue shall be correspondingly varied from the date of each such variation.

(b) In extreme cases, the Contractor may be entitled to treat the Employer's behaviour as repudiatory and thus determine (terminate) the Contract. Legal advice should be sought before taking such a draconian step (see pp.340–348, *Hudson's Building and Civil Engineering Contracts*, 10th edition).

(*c*) Arbitration can be embarked upon whenever there is a failure to certify (see clause 66(2)).
(*d*) Court proceedings could be started where a certified sum remains unpaid.

Chapter 4
Specification and quality

Never mind the quality, feel the width.

The Specification
It is the Specification which, for any given contract, lays down the quality and standard requirements. Clause 1(1)(*f*) defines 'Specification' as:

'Specification' means the specification referred to in the Tender and any modification thereof or addition thereto as may from time to time be furnished or approved in writing by the Engineer

The Specification should be the nub of the Contract: all the other elements in the Contract are supportive. The Engineer should, therefore, be positive and unequivocal in his statement of the object of the Investigation and what is required from it. If the Employer wants to give the Engineer only an indication of the object required (not uncommon in large organisations) and expects him to produce the Specification, an experienced Engineer should try to get the object of the Investigation amplified as fully as possible. He will then be able to proceed with his part of the Investigation in the manner shown in Figure 1 (p.7), and prepare a Specification which fulfills the requirement of the Employer.

Table 2 indicates the preferred sequence and scope of a typical Specification. It is not exhaustive and will require modification and/or expansion according to the type of ground Investigation to be undertaken.

It is not possible, within the scope of this book, to present a model Specification. Table 2 is an abridged list of the headings appropriate for a typical Specification for a single Site. Some examples of the details applicable under these headings are given. Generally the sequence should follow that of the Investigation (although the current CESMM (1976) does not follow that sequence).

Table 2. Draft outline contents of Specification

(*a*)	Preamble: object, known geology, history and site conditions
(*b*)	Site access and limitations of loading and on transportation
(*c*)	Definition of Sections if sectional completion is required
(*d*)	Water and electricity sources and drainage availability
(*e*)	Owners'/tenants' names and certificate and extent of permission to enter
(*f*)	Seasonal limitations regarding crops
(*g*)	Limitation of working days/hours
(*h*)	Noise limitations
(*i*)	Pollution of water courses and sources in surface or sub-surface catchment areas from wash–boring extrusions or excavated material
(*j*)	Method of measurement
(*k*)	Lists of 'Schedules'
(*l*)	Borehole, test, trial pit and sample positions: type and quality
(*m*)	Transport of samples
(*n*)	Security and confidentiality
(*o*)	Insurance arrangements
(*p*)	Safety precautions generally, and emphasis where contaminants are likely to be encountered
(*q*)	Transport of contaminated samples and contaminated material
(*r*)	Transport of radioactive samples and material
(*s*)	Disposal of contaminated and/or radioactive material
(*t*)	Requirements for soils, materials, chemical and radioactive material laboratories
(*u*)	Contents of testing schedules to be submitted
(*v*)	Reports required: type, sequence, content and by and for whom prepared
(*w*)	Design responsibility, if any, of the Contractor
(*x*)	Detailed description of Ancillary Works
(*y*)	Codes and standards (and relevant parts thereof) to be applied
(*z*)	Need for services of a specialist
(*aa*)	Employer's responsibility to provide materials, equipment or plant
(*bb*)	Reinstatement and maintenance requirements
(*cc*)	List of Contract Drawings and of plans

The preamble (item (*a*) on Table 2), will never be standard and must be drafted with great care. It should contain as much information about the object of the Investigation as possible:

without this the ultimate reports could be meaningless. The Employer and Engineer should provide any information they have about the geology, history and Site conditions. This will allow the Contractor to guard against problems this reveals, reduce the possibility of claims and ensure the high quality of the Investigation.

Clauses 29 and 30 of the *Conditions* identify the responsibilities of the Contractor with regard to traffic and transport, and should be borne in mind when drafting the Specification. Limitations regarding the movement of traffic on Site, or construction of access roads to accommodation and borehole/trial pit positions should be given. From this point, the onus of responsibility will rest with the Contractor under clauses 8, 29 and 30. If there are special requirements regarding siting, type and services to the Contractor's accommodation these should be specified. For example, a long-term investigation may be subject to local authority planning restrictions, limitations of supply and disposal, or to the services of statutory undertakers. The Engineer will normally have planned the required positions and type of boreholes and/or trial pits, and will have indicated, or anticipated, their depths. This information should be given, with references to the Drawings and plans. The Specification can then proceed to words similar to the following:

The Contractor's attention is drawn to clauses 8(1) and 8(2) of the *Conditions* in respect of safety in site operations generally and especially where operations in made ground are concerned where there is a possibility that fly-tipping may have incurred potential hazard from toxic and inflammable material and gases.

Boreholes of 200mm minimum diameter are to be made at positions nos 1 to 6 inclusive shown and to the minimum depths indicated on Drawing . . . .They are to be made in accordance with BS 5930: 1981 clause 18.4.

This brief specification assumes that the indigenous ground is such that normal light cable percussion boring will be suitable to the full depth of boring indicated. Should the preliminary or desk study have shown that hard rock in boulder form, or as major bedding, would be encountered, it would be necessary to invoke BS 5930/clause 18.7 and the relevant interval and type of samples.

Trial pits of not less than 3m in length and 1m in width are to be made in what is considered to be made ground at Positions Nos 7 to 10 inclusive shown and to the depths indicated on Drawing . . . . They are to be excavated by machine in accordance with BS 5930: 1981 clause 18.1 in stages not exceeding 200mm in depth (where the type of fill permits) and the base of the pit is to be shown level at each stage.

The quality and type of sampling requirements could, in appropriate cases, be expressed in the following form:

Sampling in boreholes will be executed in such a manner that the samples resulting are suitable for testing to determine the physical and strain propensity of the material recovered. Undisturbed samples will be of 100mm minimum diameter to quality Class 1 as BS 5930 clause 19.2 and taken as given in clause 19.4.2. The first in each boring will be taken at 1m below existing ground level, thereafter at each change of soil type and at marked change of consistency, succeeded by sequence at 2m intervals within the same stratum. Disturbed samples as BS 5930: 1981 clause 19.3 Classes 3 or 4 of not less than 2 kg in weight will be taken immediately before the undisturbed samples are taken.

[The soil testing laboratories should be required to exercise 'a standard of care of the samples with cleanliness approaching those (laboratories) dealing with clinical specimens' (see Chapter 2, p.30), but there would be little point if the extraction of samples was not carried out to the same standard. Clause 15(1) of the *Conditions* is also relevant in this context.]

Sampling from trial pits will include bulk disturbed, block undisturbed, groundwater and contaminant semi-liquid or viscous matter. They are to be regarded as potentially hazardous and protective clothing for operators will be deemed to be required under clauses 8, 15(1) and 19(2) of the *Conditions*.

Bulk disturbed samples of not less than 150kg in combined bagged weight will be taken at 1m intervals from ground level and be fully representative of the material at the base level from which they are taken. One block undisturbed sample of not less than 300mm equivalent cubed size (1 cubic foot approximately) will be taken from each trial pit at the position given by the Engineer during his inspection and will be in accordance with BS 5930: 1981 clause 19.9. Samples of groundwater or liquids and viscous matter of suspected contaminant nature extracted when encountered shall be not less than 1 litre in quantity and taken in sterilised sealable air-tight wide-necked glass bottles such as Kilner jars or similar approved containers.

If rock samples (cores) are required, these are covered by the *Conditions* under the clauses previously quoted for other types of sample, and the method and sample condition clauses of BS 5930: 1981 to be invoked are as follows:

(*a*) For rotary core samples, clause 19.8;
(*b*) block samples, clause 19.9;
(*c*) disturbed hand samples, clause 19.20.3;
(*d*) core extrusion and preservation, clause 19.10.5.

The general requirements for careful handling and secure storage under fully protective conditions are referred to in clauses 20(2) and of the *Conditions* (which are considered in Chapter 5, p.64).

There are three specific requirements regarding handling and storage of samples and cores within these clauses:

(*a*) the care in preparing the samples for storage;
(*b*) the method of transport to storage;
(*c*) the conditions under which they are stored.

All of these should be included in the Specification by invoking the relevant clause of BS 5930: 1981 and extended for the particular Investigation, where necessary.

On Site, samples of essential material, particularly 'undisturbed' samples, are often not treated with the care necessary to ensure accurate results from the laboratory tests. The Specification can usefully state:

On recovery from the boring and detachment from the rods the samples will be labelled, sealed, capped and stored in accordance with BS 3950: 1981 clause 19.10.

If samples and cores are to be stored in any particular way, in any particular location, or for any period longer than the period of 28 days after the Investigation referred to in clause 20(2) of the *Conditions*, these special requirements should be specified.

Testing will be carried out both *in situ*, that is executed in the borehole or trial pit, and also in the laboratories. It is included in 'Schedules' under clause 1(1)(*h*) of the *Conditions*.

Schedules means the schedules and lists of Site Operations Laboratory Testing and other requirements referred to in the Specification

This suggests that *in situ* testing is basically a Site Operation. This is the case when the test requires only the services of those operators engaged in boring or excavation and sampling, such as a Standard Penetration Test (BS 5930: 1981 clause 21.2 and

BS 1377: 1977, test 19). However, when *in situ* tests require the services of specialist operators such as a Vane Test, Permeability Test (BS 5930: 1981 clauses 21.3 and 21.4), or those for the detection of contaminants, they could be regarded as outside the normal sphere of Site Operations. For this reason, it is advisable that a section of the Specification is clearly headed '*In situ* tests' and the tests itemised. Laboratory tests should be treated in the same way and in both sections contaminant tests should be sub-sectioned to emphasise the specialised nature and potential hazards connected with them.

Clause 1(1)(*h*) of the *Conditions* envisages that 'schedules and lists of Site Operations Laboratory Testing and other requirements' will be given or referred to in the Specification. The Site Operations and Laboratory Testing required will be described and the sequence required will be indicated. Criteria for acceptability of laboratories should be laid down. Often a schedule can describe the form in which a boring log, test schedule or test results can be presented. The dayworks schedule should be identified in the Specification, although it is commonly to be found at the end of the Bills of Quantity. Although the FCEC standard schedule is available for use (see Appendix 2), consideration should be given to the use of the somewhat modified schedule discussed by Clapham and Hughes (*Civil Engineering*, 1983, see Appendix 3).

There should be detailed descriptions of the reports required. An explanation of their required contents and extent should be provided, and their form can also be identified, for particular sections even to the extent of illustrating the layout required. Reports are considered in Chapter 6.

Quality
The Contractor's obligation is to carry out the Investigation in accordance with all the Contract documents. Quality and the standards are determined:

(*a*) by the express terms of the Contract;
(*b*) by codes of practice (if incorporated); and
(*c*) to the extent that the Contract is silent in any particular respect, by implied terms that the materials shall be of merchantable quality and that the work should be executed in a proper workmanlike manner.

Clauses 36(1) and (3) lay down express requirements as to quality:

(1) All materials supplied by the Contractor and workmanship shall be of the respective kinds described in the Contract and in accordance with the Engineer's instructions and shall be subjected from time to time to such tests as the Engineer may direct at the place of manufacture or fabrication or on the Site or such other places as may be specified in the Contract. The Contractor shall provide such assistance instruments machines labour and materials as are normally required for examining measuring and testing any Ancillary Works and the quality weight or quantity of any materials used and shall supply for testing samples of materials before incorporation in the Ancillary Works or use in the Site Operations as may be selected and required by the Engineer.
(3) All Equipment shall be of the respective type and kind suitable for the proper execution of the Investigation in accordance with the Engineer's instructions and shall be calibrated and subjected to such calibration tests as may be necessary to ensure performance in accordance with the requirements of the Contract.

Failure on the part of the Contractor to comply with contractual requirements regarding standard and quality put him in breach of contract for which he is liable for damages to the Employer.

To ensure proper completion of the Investigation, the Engineer on behalf of his Employer is provided by the *Conditions* with the widest battery of powers:

(*a*) to supervise;
(*b*) to instruct;
(*c*) to remove incompetent personnel;
(*d*) to require remedial works;
(*e*) to determine.

These powers will now be considered.

(a) Power to supervise
Supervision of a ground Investigation by the Engineer, as directed by clauses 2(1) to 2(3) quoted below, will vary from site visits to determine and record progress under the several items of the Bill of Quantities and/or Schedule of Dayworks, to continuous supervision by the Engineer's Representative or a person named in the Contract documents.

(1) The functions of the Engineer's Representative are to watch and supervise the Investigation. He shall have no authority to relieve the Contractor of any of his duties or obligations under the Contract nor except as expressly provided hereunder to order any work involving delay or any extra payment by the Employer nor to make any variation of or in the Investigation.

(2) The Engineer or the Engineer's Representative may appoint any number of persons to assist the Engineer's Representative in the exercise of his functions under sub-clause (1) of this Clause. He shall notify to the Contractor the names and functions of such persons. The said assistants shall have no power to issue any instructions to the Contractor save insofar as such instructions may be necessary to enable them to discharge their functions and to secure their acceptance of methods materials workmanship or Ancillary Works as being in accordance with the Specification Drawings and Schedules and any instructions given by any of them for those purposes shall be deemed to have been given by the Engineer's Representative.

(3) The Engineer may from time to time in writing authorise the Engineer's Representative or any other person responsible to the Engineer to act on behalf of the Engineer either generally in respect of the Contract or specifically in respect of particular Clauses of these Conditions of Contract and any act of any such person within the scope of this authority shall for the purposes of the contract constitute an act of the Engineer. Prior notice in writing of any such authorisation shall be given by the Engineer to the Contractor. Such authorisation shall continue in force until such time as the Engineer shall notify the Contractor in writing that the same is determined. Provided that such authorisation shall not be given in respect of any decision to be taken or certificate to be issued under Clauses 12(3) 44 48 60(3) 61 63 and 66.

The Engineer may think that these are the sole duties and responsibilities of the person, or persons, concerned but that is not the case: if there is any divergence from the requirements of the Contract, the method of working or safety precautions, the person concerned could be legitimately criticised for non-fulfilment of his duties under the Contract, or of professional negligence in tort, should he not have observed or, having observed, failed to report or correct the malpractice as soon as possible. A minor, but important, example would be connecting the steel wire rope of a light cable percussion rig to the working tool by means of simple knots rather than by shackle, eye and clamps or splicing. This actually happened and as a result the cable was severed and the tool was lost at the bottom of the boring. Subsequently the ground settled and a structure collapsed as a result of misguided attempts to recover the tool.

As mentioned on p.60 the Engineer or the Engineer's Representative might, in normal methods of working, be tempted to interfere with the Contractor's progress. Unwarranted interference may result in a legitimate claim against the Employer by the Contractor for damages. It is essential that the requirements of the codes of practice relating to the methods of execution chosen by the Contractor, and approved by the Engineer, are observed. This applies particularly to sampling and *in situ* testing where slipshod methods would result in unrepresentative specimens and inaccurate results.

If the extent of the supervisory function is going to be unusual, it should be set out in the Specification. For instance, if the materials encountered are to be subjected to a forensic examination on an inch-by-inch basis, this must be specified, since otherwise it might not reasonably be inferred as part of the Contractor's obligation under clause 8(1).

Whatever else is specified, clause 37 provides the Engineer with wide powers of access not only to the Site but elsewhere:

> The Engineer and any person authorised by him shall at all times have access to the Site and to the Site Operations and Laboratory Testing and to samples wherever stored and to all laboratories workshops and places where work is being prepared or carried out whence materials manufactured articles and machinery are being obtained for the Investigation and the Contractor shall afford every facility for and every assistance in or in obtaining the right to such access.

Under this clause the co-operation of the Contractor can be demanded to enable the Engineer or his delegates to monitor every stage of the Investigation.

(b) Power to instruct
Clause 13(1) enables the Engineer to issue any instructions or directions to the Contractor on any matter connected with the Contract. Although there are many clauses which allow specific types of instruction to be given (e.g. clause 51, variations), clause 13(1) gives a residual power to the Engineer to issue instructions (even though they are not mentioned in the Contract). Clause 13(1) instructions can postpone the need for an argument about whether or not the instruction is a variation.

(c) Power to remove incompetent personnel
A ground Investigation will usually involve a higher propor-

tion of qualified and skilled personnel than a normal civil engineering project. It is crucial, accordingly, that suitable persons are used; it is not uncommon for the Specification to identify the skill or qualification requirements for particular jobs. The Engineer is, by clause 16:

at liberty to object to and require the Contractor to remove from the Investigation any person employed by the Contractor in or about the Investigation who in the opinion of the Engineer misconducts himself or is incompetent or negligent in the performance of his duties or fails to conform with any particular provisions with regard to safety which may be set out in the Specification or persists in any conduct which is prejudicial to safety or health and such persons shall not again be employed upon the Investigation without the permission of the Engineer.

'Misconduct' could mean any conduct which seriously prejudices the Investigation or the contractual administration of the Investigation. For instance, a Contractor's employee who sold the story of a highly confidential Site Investigation to a newspaper might 'misconduct himself'; an employee who vandalised Equipment, or partly-built Ancillary Works, or an employee who publicly defamed the Engineer, would 'misconduct himself'. The Engineer's powers under Clause 16 are limited to the Contractor's employees.

(d) Power to require remedial works
The Engineer ultimately has power to have defective materials or work put right. Clause 39(1) states:

The Engineer shall during the progress of the Site Operations have power to order in writing:
(a) the removal from the Site within such time or times as may be specified in the order of any materials which in the opinion of the Engineer are not in accordance with the Contract;
(b) the substitution of proper and suitable materials; and
(c) the rectification or the removal and proper re-execution (notwithstanding any previous test thereof or interim payment therefor) of any work which in respect of materials or workmanship is not in the opinion of the Engineer in accordance with the Contract.

This clause is wide enough to allow re-execution of sampling, or testing, made necessary by incompetent or bad workmanship on the part of the Contractor: the order for re-execution must be given before completion of the Site Operations. The Employer is protected against the failure of the Contractor to re-execute by clause 39(2).

> In case of default on the part of the Contractor in carrying out such order the Employer shall be entitled to employ and pay other persons to carry out the same and all expenses consequent thereon or incidental thereto shall be borne by the Contractor and shall be recoverable from him by the Employer or may be deducted by the Employer from any monies due or which may become due to the Contractor.

Clause 39(1) is complemented by clause 36(4), which states:

> The cost of making any test or calibration pursuant to this clause shall be borne by the Contractor if such tests or calibrations are clearly intended by or provided for in the Contract and in the case of a test to ascertain whether the design of any finished or partially finished work is appropriate for the purposes which it was intended to fulfil if it is particularised in the Specification or Bill of Quantities in sufficient detail to enable the Contractor to have priced or allowed for the same in his Tender. If any test or calibration is ordered by the Engineer which is either not so intended by or provided for or is not so particularised then the cost of such test or calibration shall be borne by the Contractor if the test or calibration shows the Ancillary Works Equipment workmanship or materials not to be in accordance with the provision of the Contract or the Engineer's instructions but otherwise by the Employer.

The Engineer may order the Contractor to carry out tests to ascertain whether the Contract has been complied with.

There is one gap in the powers of the Engineer: there is no expressed contractual power to require the re-execution of defective Laboratory testing work after completion of the Site Operations. The Engineer would probably have power to order this under clause 13. In any event the Employer would have a right to damages.

(c) Power to determine

The final power, where all else has failed, is for the Employer to expel the Contractor from the Site, if the Engineer certifies in writing under clause 63(1) that the Contractor:

> (c)   has failed to remove goods or materials from the Site or to rectify or to pull down and replace work for 14 days after receiving from the Engineer written notice that the said goods materials or work have been condemned and rejected by the Engineer; or
>
> (d)   despite previous warning by the Engineer in writing is failing to

proceed with the Investigation with due diligence or is otherwise persistently or fundamentally in breach of his obligations under the Contract; or

(*e*) has to the detriment of good workmanship or in defiance of the Engineer's instruction to the contrary sub-let any part of the Contract;

The procedure in clause 63(1) must be followed to the letter to ensure that the expulsion is valid.

Defects can be required to be put right at any time up to 14 days from the expiry of the Period of Maintenance (see Chapter 6, p.76).

Chapter 5
Care, security, superintendence and insurance

A dusk misfeatured messenger
No other than the angel of this life
Whose care is lest man see too much at once.
Robert Browning

Care of the Investigation

Clauses 19, 20 and 22 of the *Conditions* apportion the responsibilities of the parties for specified contingencies. For individual sites and investigations, special provisions may have to be agreed between the parties at Contract stage.

During the Site Operations, the Contractor is primarily responsible for the care of them, with the exception of 'Excepted Risks'. Clause 20(1) provides for this:

The Contractor shall take full responsibility for the care of the Site Operations from the date of the commencement thereof until 14 days after the Engineer shall have issued a Certificate of Completion for the whole of the Site Operations pursuant to Clause 48. Provided that if the Engineer shall issue a Certificate of Completion in respect of any Section or part of the Site Operations before he shall issue a Certificate of Completion in respect of the whole of the Site Operations the Contractor shall cease to be responsible for the care of the Site Operations and the Ancillary Works comprised within that Section or part 14 days after the Engineer shall have issued the Certificate of Completion in respect of that Section or part and the responsibility for the care of the Site relating to that Section or part and of any Ancillary Works included in the Section or part shall thereupon pass to the Employer. Provided further that the Contractor shall take full responsibility for the care of any outstanding work which he shall have undertaken to finish during the Period of Maintenance until such outstanding work is complete.

This responsibility, which may arise irrespective of fault on the part of the Contractor, is a heavy one. For instance, if heavy rain or flooding causes an unlined borehole or a trial pit to collapse, the Contractor whether he is insured or not, is contractually liable to the Employer; if vandals destroy or damage boring equipment, the Contractor must look to his own purse (or insurers) for recompense and not to the Employer. The fact that events for which contractually the Contractor is responsible may entitle him to an extension of time under clause 44 does not detract from the onus of clause 20(1). The Excepted Risks are considered below.

In a ground Investigation, responsibility continues away from the Site itself whilst cores and samples are being transported to and tested at laboratories. Clause 20(2) imposes this responsibility upon the Contractor, save where Excepted Risks occur:

The Contractor shall unless it is otherwise provided for in the Contract take full responsibility for the care and storage of the samples and cores obtained from the investigation at his cost, until 28 days after the issue of the Report to the Engineer or in the case of a phased investigation the relevant section of the Report. After the said period the Contractor shall give the Engineer 14 days written notice of his intention to charge rental for the storage of the cores and samples. The Engineer shall then either give instructions for the immediate disposal of the samples and cores at the Contractor's expense or state his storage or other requirements giving an indication of his programme. The rental charged for the cores and samples stored shall commence after the expiry of the said 14 days notice and shall continue until the Engineer gives disposal instructions.

A traffic accident which destroys samples in transit from the Site to the laboratory is the Contractor's risk; a fire which destroys the laboratory, ruins the cores in storage there and burns all original and copy drilling logs, whilst unfortunate for the Contractor, will not attract any contractual benefit to the Contractor.

The Excepted Risks are set out in clause 20(4):

The 'Excepted Risks' are riot war invasion act of foreign enemies hostilities (whether war be declared or not) civil war rebellion revolution insurrection or military or usurped power ionising radiations or contamination by radioactivity from any nuclear fuel or from any nuclear waste from the combustion of nuclear fuel radioactive toxic explosive or other hazardous properties of any explosive nuclear assembly or nuclear component thereof pressure waves caused by aircraft or other aerial devices travelling at sonic or supersonic speeds or a cause due to use or occupation by the Employer his agents servants or other contractors (not being employed by the Contractor) of any part of the Site Operations or to fault defect error or omission in the design of the Investigation (other than a design provided by the Contractor pursuant to his obligations under the Contract).

If the Site to be investigated is one which is known or suspected to be contaminated with radioactive materials or gases, amendments may have to be made to this standard term.

The most commonly encountered Excepted Risk is 'fault defect error or omission in the (Engineer's/Employer's) design of the Investigation'. Shortcomings in the design of any permanent works to be built at the Site are easily comprehensible: if a reinforced concrete monitoring station is to be built and it fails, for example, because of the inadequacy of the specified reinforcement, the Contractor is not responsible. In ground investigations, a defect in the design of the Investigation need not arise through fault on the part of the 'designer'. If the inevitable effect of specified boring is to ensure that adjacent boring either cannot be carried out or is damaged, the risk rests with the Employer.

Clause 20(3) lays down the consequences of any damage, loss or injury which happens to the Site Operations, the samples and cores, or to the Report:

> In case any damage loss or injury from any cause whatsoever (save and except the Excepted Risks as defined in sub-clause (4) of this Clause) shall happen to the Site Operations the samples and cores or the Report or any part thereof while the Contractor shall be responsible for the care thereof the Contractor shall at his own cost repair make good or replace the same so that at completion the Ancillary Works shall be in good order and condition and the Investigation shall be in conformity in every respect with the requirements of the Contract and the Engineer's instructions. To the extent that any such damage loss or injury arises from any of the Excepted Risks the Contractor shall if required by the Engineer repair make good or replace the same as aforesaid at the expense of the Employer. The Contractor shall also be liable for any damage to the Ancillary Works occasioned by him in the course of any operations carried out by him for the purpose of completing any outstanding work or for complying with his obligations under Clauses 49 and 50.

Most, if not all, of the risks assumed by the Contractor and the Employer not only are insurable, but must be insured against (see p.71).

It must be remembered that clause 20 regulates risk simply between the Employer and the Contractor. It will not necessarily provide any defence to a claim brought by a third party.

Safety and security

Primarily, the Contractor is responsible for the safety and security aspects of the Site Operations whilst they are being executed. Clause 19(1) states:

The Contractor shall throughout the progress of the Site Operations have full regard for the safety of all persons entitled to be upon the Site and shall keep the Site (so far as the same is under his control) and the Site Operations (except insofar as the same consist of Ancillary Works covered by a Certificate of Completion or which have been delivered up to the Employer) in an orderly state appropriate to the avoidance of danger to such persons and shall *inter alia* in connection with the Site Operations provide and maintain at his own cost all lights guards fencing warning signs and watching when and where necessary or required by the Engineer or by any competent statutory or other authority for the protection of the Site Operations or for the safety and convenience of the public or others.

This clause, coupled with clause 26, requires the Contractor to carry out the Site Operations in a safe manner and to comply with all appropriate health and safety requirements (both statutory and otherwise).

The clause is of particular importance to the investigation of contaminated Sites and is discussed in Chapter 8 (p.111), in relation to hazards and protective clothing. In this respect, it may be appropriate for either Contractor or Engineer to seek advice from a specialist as to the type of preventive and protective temporary structures necessary. A basic requirement is that the Site should be kept tidy: all apparatus used in the operations should be cleaned and laid out neatly for re-use, samples taken to a particular point for sealing and stacked on prepared boards or 'clean' areas prior to removal to the storage place. Supervision by the Engineer or Engineer's Representative can ensure that these requirements are maintained on Site at all times.

Clause 19(2) contains a limited exception to clause 19(1) in the case of the Employer's workmen and other contractors:

If under Clause 31 the Employer shall carry out work on the Site with his own workmen he shall in respect of such work:
(a) have full regard to the safety of all persons entitled to be upon the Site; and
(b) keep the Site in an orderly state appropriate to the avoidance of danger to such persons.

If under Clause 31 the Employer shall employ other contractors on the Site he shall require them to have the same regard for safety and avoidance of danger.

It should be noted that the Employer's obligations in this case are not as extensive as those of the Contractor under clause 19(1).

Clauses 19(1), 19(2), 20 and 22 should be carefully followed in Investigations involving contaminated land, from the commencement of Site Operations, during the obtaining, carriage, storage and testing of samples and in the use of tools and plant. Physical contact with contaminants on Site may result in immediate visible injury or may damage the health of the Site operators in a way which may not be evident until some time later. The parties should bear in mind that injury or ill-health may be caused by contact with, or inhalation of, samples in storage, specimens being tested or from material adhering to tools, apparatus and plant which has not been cleaned before leaving the Site. The Contractor may also be liable for the release into the atmosphere of toxic or other injurious gases which may affect people off Site as well as those on Site.

Pollution of underlying strata may be caused by the boring passing through the contaminated material and penetrating an aquifer leading to an outlet of a potable water supply or watercourse. Similarly, water discharged from flush boring may pass through contaminated land before disposal and could be a source of pollution. The Contractor could be required to neutralise the contaminant before disposing of the water off Site. It should be noted that permission of the water authority must be obtained before boring in a water collection or source area.

The Engineer's powers, in clause 19, to procure or secure safe working conditions on the Site must be exercised with care: he must achieve safety without assuming a responsibility which is primarily the Contractor's. The Engineer may be liable to a third party for negligence if, having discovered a critical inadequacy in the Contractor's safety precautions, he does nothing or compounds the inadequacy. If he instructs the Contractor to provide a particular precaution, the Contractor must comply; if compliance with the instruction causes damage or injury to third parties, the Employer and Engineer may be liable to that third party.

Confidentiality and security

Persons or bodies whose interests are in conflict with those of the Employer may obtain valuable information free by knowledgeable and trained observation of Site Operations even from some distance. This is dealt with by clause 18 of the *Conditions*:

The Contractor shall at all times keep confidential between himself the Engineer and the Employer all information obtained from the carrying out of the Contract and no information relating thereto shall be communicated in any form to any person or body other than those named in written authorisation given by the Employer or the Engineer. The Contractor shall so far as is reasonably practicable prevent the entry of unauthorised persons to the Site to examine the work in progress and safeguard and secure samples taken from the boring or trial pits and records against unauthorised examination.

Where such interests are at stake it would be prudent for the Engineer to emphasise this in the preamble to the Specification. He should also consider the inclusion in the Specification and Bill of Quantities of an item for the provision of temporary screening for individual boring positions. This screening should block the operations from view without impeding the progress of the work.

Clause 18 is very difficult to enforce. A breach of confidence by the Contractor may not be capable of compensation by money. The Courts would usually act quickly on an application for an injunction to restrain the Contractor from breach of Clause 18.

Damage to persons and property

On certain ground investigations damage, or loss, will be caused to persons or property (land or things). Clause 22 identifies which party is to bear ultimate responsibility for such loss or damage:

(1) The Contractor shall (except if and so far as the Contract otherwise provides) indemnify and keep indemnified the Employer against all losses and claims for injuries or damage to any person or property whatsoever (other than the Investigation for which insurance is required under Clause 21 but including surface or other damage to land being the Site suffered by any persons in beneficial occupation of such land) which may arise out of or

69

in consequence of the Investigation and against all claims demands proceedings damages costs charges and expenses whatsoever in respect thereof or in relation thereto. Provided always that:

(a)   the Contractor's liability to indemnify the Employer as aforesaid shall be reduced proportionately to the extent that the act or neglect of the Employer his servants or agents may have contributed to the said loss injury or damage;

(b)   nothing herein contained shall be deemed to render the Contractor liable for or in respect of or to indemnify the Employer against any compensation or damages for or with respect to:
(i) damage to crops being on the Site (save insofar as possession has not been given to the Contractor);
(ii) the use or occupation of land (which has been provided by the Employer) for the purpose of carrying out Site Operations (including consequent losses of crops) or interference whether temporary or permanent with any right of way light air or water or other easement or quasi easement which are the unavoidable result of the Site Operations carried out in accordance with the Contract;
(iii) the right of the Employer to have the Site Operations or any part thereof carried out on over under in or through any land;
(iv) damage which is the unavoidable result of the Site Operations in accordance with the Contract;
(v) injuries or damage to persons or property resulting from any act or neglect or breach of statutory duty done or committed by the Engineer or the Employer his agents servants or other contractors (not being employed by the Contractor) or for or in respect of any claims demands proceedings damages costs charges and expenses in respect thereof or in relation thereto.

(2) The Employer will save harmless and indemnify the Contractor from and against all claims demands proceedings damages costs charges and expenses in respect of the matters referred to in the proviso to sub-clause (1) of this Clause. Provided always that the Employer's liability to indemnify the Contractor under paragraph (v) of proviso (b) to sub-clause (1) of this Clause shall be reduced proportionately to the extent that the act or neglect of the Contractor or his sub-contractors servants or agents may have contributed to the said injury or damage.

It should be emphasised that any loss or damage caused by Excepted Risk is the responsibility of the Employer (clauses 20(3) and 20(4)).

The contingencies set out in clauses 22(1)(b)(ii) and (iv) are especially applicable to ground investigation. If it is necessary to locate a borehole in a stream, and the unavoidable consequence is the pollution of that stream, the Employer must contractually bear the responsibility. If it is specified that a heavy rig is to be used in a location where inevitable noise will cause nuisance to a third party, the Contractor should be fully reimbursed by the Employer. If a trial hole has to be located in a particular position and it undermines or affects the support of an adjacent building, the Employer must pay for the consequences.

Insurance

The Contractor is required by clauses 21 and 23 respectively to obtain adequate insurance in respect of those matters for which he is liable under clauses 20 and 22 respectively. The insurance under clause 21 must be in the joint names of the Employer and the Contractor (so that both parties may proceed against the insurance company) but the insurance under clause 23 need not be in joint names. The insurance clauses do not prevent either party from taking out additional insurance nor does it prejudice them if they have done so.

Liability to third parties

Liability at common law to third parties (in nuisance, trespass or negligence, for example) is largely unaffected by the care, indemnity and insurance clauses. Provided that both Employer and Contractor remain solvent or fully insured against third party risks, the ultimate financial risks are contractually apportioned. Problems do arise where one party becomes insolvent or disappears. The legal responsibility of the designers and supervisors has been clarified by recent case law.

It is now settled law that a Contractor, Employer, Engineer or sub-contractor owes a duty in tort to all people who may forseeably be affected by their acts or omissions. Thus, where a Contractor is required by the Engineer to do something, as part of the Site Operations, which he knows (and the Engineer should know) would cause damage to a neighbour, both the Contractor and the Engineer would be liable to that neighbour for negligence. An Employer who employs the Contractor to dig a trial pit which will undermine a neighbour's building or reduce the stability of his land is liable in nuisance to that neighbour; the Contractor, albeit innocent (and not even negligent) may also be liable to the neighbour.

Superintendence

There are two anticipated consequences of good superintendence by the Contractor: quality (considered in Chapters 2 and 4) and safety. The clauses in the Contract which deal with superintendence can, and ideally should, be complemented by BS 5930: 1981.

Clause 15, sub-clause (1) of BS 5930: 1981 relating to 'Personnel for Ground Investigation' states:

Introduction:
In view of the importance of ground investigation as a fundamental preliminary to the proper design and efficient economical construction of all civil engineering and building works, it is essential that the personnel involved in the investigation should have appropriate specialised knowledge and experience, and be familiar with the purpose of the work.

Sub-clauses 15(2) to 15(8) of BS 5930: 1981, deal in turn with the qualifications, experience, skills and application required of those persons jointly and individually responsible for the several stages of the Investigation, including the preparation of the Report. BS 5930: 1981 does not say whether these personnel are the Engineer's or the Contractor's; but it states quite firmly that the criteria must be applied to those involved in the Investigation.

Where geotechnical and ground investigation expertise is not available to the Employer it is recommended that these sub-clauses, together with those dealing with specialist personnel for operations in contaminated ground should be invoked in the Specification in support of clauses 15(1) and 15(2) of the *Conditions*. Specialist personnel for contaminated land are dealt with generally in clause 13.4 and Appendix E of BS 5930. Specific detail is given in the draft British Standard code of practice on *The identification and investigation of contaminated land* which was circulated for public comment in December 1983 under reference number EPC/47. It has not yet been decided whether this will be published as a separate code, incorporated in BS 5930 or as BS 5930 Part 2.

Sub-clause 15(1) states:

The Contractor shall give or provide all necessary superintendence during the execution of the Investigation. Such superintendence shall be given by sufficient persons being suitably qualified having adequate experience and knowledge of the operations to be carried out (including the methods and techniques required the hazards likely to be encountered and the prevention of accidents) as may be requisite for the satisfactory execution of the Investigation. The Contractor shall be responsible for the safety of all operations.

It will be noted that this clause, being part of the General Obligations, does not identify the degree of expertise as does BS 5930: 1981.

This sub-clause effectively requires the Contractor to provide suitable superintendent personnel for the Investigation at all stages. Even if proper superintendence is provided, it does not relieve the Contractor of responsibility for failures to comply with the Contract.

The importance of the sub-clause on a ground investigation should not be underestimated. For example, if the soils encountered by the operator are incorrectly described on the Site log recurring errors will result. Those who are responsible for the initial interpretation of the findings may be misled and the decisions in the preliminary recommendations to the Employer may be incorrect, and laboratory tests which are inapplicable to the specimen may be carried out. The Contractor should make a habit of examining the Site log against the sample on Site. Close supervision is important in forensic investigations where the reasons for failure of a structure or the stability of a soil mass may be revealed during the investigation. In such investigations, the Contract assumes that the Contractor's staff are experienced and able to detect significant changes in the classification and condition of soil or rock encountered, and that they will vary the type, or sequence, of sampling and/or *in situ* testing accordingly. Variations such as these, which are the result of observation, can be initiated by the Engineer's Representative or by the Contractor, as both should be aware of the mode of failure of the structure, and thus competent to suggest operations to determine the reason for that failure. The position, type and dimensions of a slip plane or zone, for example, may vary considerably depending on the other factors involved. Superintendence of the handling, recording, transport and storage of samples and the manner in which specimens from the samples are prepared and tested in the laboratory is also generally necessary.

Superintendence of the investigation of contaminated land may require the continuous attendance on Site of the directing specialist if the Contractor lacks experience of such investigations. This is because on contaminated land errors in methods of working and/or the lack of adequate, or instant, safety precautions when toxic or radioactive materials are encountered could lead to disastrous consequences.

Superintendence by the Contractor requires liaison and agreement to ensure the proper performance of the Contract. Clause 15(1) is complemented by clause 16:

> The Contractor shall employ or cause to be employed in and about the execution of the Investigation and in the superintendence thereof only such persons as are careful skilled and experienced in their several trades and callings and the Engineer shall be at liberty to object to and require the Contractor to remove from the Investigation any person employed by the Contractor in or about the Investigation who in the opinion of the Engineer misconducts himself or is incompetent or negligent in the performance of his duties or fails to conform with any particular provisions with regard to safety which may be set out in the Specification or persists in any conduct which is prejudicial to safety or health and such persons shall not again be employed upon the Investigation without the permission of the Engineer.

This clause applies to the Contractor's superintendent, other employed staff, qualified and unqualified persons, and also, in part, to the staff of sub-contractors and the laboratories used by the Contractor.

Clauses 15(2) and 15(3) of the *Conditions* are more in line with BS 5930: 1981; sub-clause 15(3) allows for the services of a specialist (if specified) 'available on site or elsewhere'. Specialists with expertise other than geotechnical are included by the words 'other technical and advisory services':

> (2) The Contractor shall provide a competent suitably qualified and authorised agent or representative approved of in writing by the Engineer (which approval may at any time be withdrawn). Such authorised agent or representative shall be in full charge of the Investigation and shall receive on behalf of the Contractor directions and instructions from the Engineer or (subject to the limitations of Clause 2) the Engineer's Representative. Such agent or representative may appoint a person who shall be constantly on the Site during the Site Operations and shall receive instructions relating to the Site Operations.
> (3) If in addition to the superintendence in accordance with sub-clauses (1) and (2) of this Clause the Contract shall require or the Engineer direct the Contractor to make available on the Site or elsewhere the services of suitably qualified persons for description of soils and rocks logging of trial pits execution of geological and geotechnical appraisals other technical and advisory services and the preparation of technical reports the extent and scope of the service required shall be specified in the Contract.

From this it follows that in most ground Investigations the

Contractor's agent should be qualified and have some experience or, if he is unqualified, he should be very experienced. He should, for instance, be competent to make the executive and engineering decision to continue with drilling without further instruction, in anticipation of a variation under clause 13(4).

Chapter 6
Completion and reporting

*There are two things which I am confident that I can do
very well: one is an introduction to any literary work. . .
the other is a conclusion, showing from various causes
why the execution has not been equal to what the author
promised himself and to the public.*
Samuel Johnson

Completion

The programme which the Contractor has presented and the
completion of several Sections of the Ground Investigation
within the agreed periods, where applicable (as stated in the
Appendix to the Form of Tender and covered by clause 14), are
discussed in Chapter 2, p.31. 'Completion' will be judged
against this yardstick and will be subject to any variations
which have been made. It is, therefore, advisable that Certificates of Completion are specific to particular phases of the
Site Operations, for example, the completion of a single deep
borehole involving expensive drilling, *in situ* testing and
sampling, or the completion of a Section as delineated in the
Drawings and/or identified in the Specification and Appendix
to the Form of Tender, or, on a smaller Investigation, the
completion of Laboratory Testing. Exactly when the Engineer
presents Certificates of Completion will depend upon the
extent of·the Investigation and also upon the financial ability
of the Contractor to carry his expenditure on the more
complex operations or on a particular Section.

This requirement is given by clause 48. Sub-clause (3) deals
with the completion of parts of the Investigation. Clause 48:

(1) When the Contractor shall consider that the whole of the Investigation
has been substantially completed in accordance with the requirements of the
Contract he may give notice to that effect to the Engineer or to the
Engineer's Representative. In all cases such notice shall be in writing and
shall be deemed to be a request by the Contractor for the Engineer to issue a
Certificate of Completion in respect of the Investigation and the Engineer
shall within 14 days of the delivery of such notice either issue to the

Contractor (with a copy to the Employer) a Certificate of Completion stating the date on which in his opinion the Investigation was substantially completed in accordance with the Contract or else give instructions in writing to the Contractor specifying all the work which in the Engineer's opinion requires to be done before the issue of such Certificate. If the Engineer shall give such instructions the Contractor shall be entitled to receive such Certificate of Completion within 21 days of completion to the satisfaction of the Engineer of the work specified in the said instructions.

(3) If the Engineer shall be of the opinion that any part of the Investigation shall have been substantially completed in accordance with the Contract he may issue a Certificate of Completion in respect of that part of the Investigation before completion of the whole of the Investigation. Where such part shall comprise Site Operations upon the issue of such certificate the Contractor shall be deemed to have undertaken to complete any outstanding work in that part of the Site Operations during the Period of Maintenance.

(4) Provided always that a Certificate of Completion given in respect of any Section or part of the Investigation before completion of the whole shall not be deemed to certify completion of any ground or surfaces requiring reinstatement unless such certificate shall expressly so state.

Where complex investigations are concerned, involving a long delay between completion of the Site Operations with its related Laboratory Testing and the completion of the final Report, it may not be acceptable to regard completion as inclusive of the Report. In such a case, it would be prudent to establish an agreed ratio of value of the work involved in preparing the draft and final reports against that of the Site Operations and Laboratory testing. This would mean that payment on completion subject to clause 47 would apply separately to the Report. In this way the liquidated damages for delay, in the Appendix to the Form of Tender and as given by sub-clause 47(1)(*a*), could apply separately to the Report.

Given this approach, the Appendix to the Form of Tender could, in certain cases, be modified to include reference to draft and final Reports itemised under clauses 48(1) and (2) as for the Sections.

Such modifications and their effectiveness will depend upon the specified manner and sequence of reporting, because much of the information contained in the final Report will have been given to the Employer in the form of Site logs and interim or sequential Reports; they may have been used by him in formulating a decision involving financial or practical consequences. In the case of site Investigation requiring this sequence of reporting, only the Contractor's final refined opinions and recommendations will not be available, and the monetary value of the final Report may be considered reduced accordingly.

Completion of Site Operations as a whole includes reinstatement of boreholes, trial pits, removal of access roads, temporary structures and fencing and, where applicable, the restoration of the Site to the condition existing before the Investigation commenced (although for completion of parts or Sections, such reinstatement is not required). The Engineer should consider carefully before removing temporary structures on enclosed sites, or on Sites adjacent to existing structures, where the temporary structures have been erected to safeguard existing structures. It is possible that, when the ground returns to its natural equilibrium and stability, Site Operations may have weakened, or may weaken, adjacent land or structures rendering the removal of the temporary structures hazardous. This should have been foreseen by the experienced Engineer when preparing the Specification and also by the Contractor during the pre-tender inspection of the Site; such temporary structures could be regarded (if so specified) as Ancillary Works, and left in place, even though they were intended as temporary works for the Site Operations. This could apply to temporary works below and above ground level, for example, sheet piling to trial pits and strutting where vibratory or percussive effects can be expected from the Site Operations.

Clause 33 of the *Conditions* directs that the Site should be cleared on completion:

> On the completion of the Site Operations the Contractor shall clear away and remove from the Site all Equipment surplus material and rubbish of every kind and leave the whole of the Site and any Ancillary Works clean and in a workmanlike condition to the satisfaction of the Engineer.

Period of Maintenance

The period of maintenance is specified in the Appendix to the Form of Tender and is defined by clause 49, sub-clause (1).

It should be noted that the maintenance obligations apply only to defects in the Site Operations. They do not apply to the other aspects of the Investigation (such as Laboratory Testing). Clause 49 states:

(1) In these Conditions the expression 'Period of Maintenance' shall mean

78

the period of maintenance named in the Appendix to the Form of Tender calculated from the date of completion of the Site Operations or any Section or part thereof certified by the Engineer in accordance with Clause 48 as the case may be.

(2) To the intent that the Site and the Ancillary Works and any Section or part thereof shall at or as soon as is practicable after the expiration of the relevant Period of Maintenance be delivered up to the Employer in the condition required by the Contract (fair wear and tear excepted) to the satisfaction of the Engineer the Contractor shall finish the work (if any) outstanding at the date of completion as certified under Clause 48 as soon as may be practicable after such date and shall execute all such work of repair amendment reconstruction rectification and making good of defects imperfections shrinkages damage subsidence of backfill or other faults resulting from the execution of the Site Operations as may during the Period of Maintenance or within 14 days after its expiration be required of the Contractor in writing by the Engineer as a result of an inspection made by or on behalf of the Engineer prior to its expiration.

(3) All such work shall be carried out by the Contractor at his own expense if the necessity thereof shall in the opinion of the Engineer be due to the use of materials or workmanship not in accordance with the Contract or to neglect or failure on the part of the Contractor to comply with any obligation expressed or implied on the Contractor's part under the Contract. If in the opinion of the Engineer such necessity shall be due to any other cause the value of such work shall be ascertained and paid for as if it were additional work.

(4) If the Contractor shall fail to do any such work as aforesaid required by the Engineer the Employer shall be entitled to carry out such work by his own workmen or by other contractors and if such work is work which the Contractor should have carried out at the Contractor's own cost the Employer shall be entitled to recover from the Contractor the cost thereof or may deduct the same from any monies due or that become due to the Contractor.

(5) Provided always that if in the course or for the purposes of the execution of the Site Operations or any part thereof any highway or other road or way shall have been broken into then notwithstanding anything herein contained:

(a) If the permanent reinstatement of such highway or other road or way is to be carried out by the appropriate Highway Authority or by some person other than the Contractor (or any sub-contractor to him) the Contractor shall at his own cost and independently of any requirements of or notice from the Engineer be responsible for the making good of any subsidence or shrinkage or other defect imperfection or fault in the temporary reinstatement of such highway or other road or way and for the execution of any necessary repair or amendment thereof from whatever cause the necessity arises until the end of the Period of Maintenance in respect of the works beneath such highway or other road or way or until the Highway Authority or other person as aforesaid shall have taken possession of the Site for the purpose of carrying out permanent reinstatement (whichever is the earlier) and shall indemnify and save harmless the Employer against and from any damage or injury to the Employer or to third parties arising out or in

79

consequence of any neglect or failure of the Contractor to comply with the foregoing obligations or any of them and against and from all claims demands proceedings damages costs charges and expenses whatsoever in respect thereof or in relation thereto. As from the end of such Period of Maintenance or the taking possession as aforesaid (whichever shall first happen) the Employer shall indemnify and save harmless the Contractor against and from any damage or injury as aforesaid arising out or in consequence of or in connection with the said permanent reinstatement or any defect imperfection or failure of or in such work of permanent reinstatement and against and from all claims demands proceedings damages costs charges and expenses whatsoever in respect thereof or in relation thereto.

(*b*) Where the Highway Authority or other person as aforesaid shall take possession of the Site as aforesaid in sections or lengths the responsibility of the Contractor under paragraph (a) of this sub-clause shall cease in regard to any such section or length at the time possession thereof is so taken but shall during the continuance of the said Period of Maintenance continue in regard to any length of which possession has not been so taken and the indemnities given by the Contractor and the Employer respectively under the said paragraph shall be construed and have effect accordingly.

Sub-clause (2) is self-explanatory, but the Engineer's attention is drawn to the requirement for work 'required of the Contractor in writing by the Engineer as a result of an inspection made by or on behalf of the Engineer prior to its expiration'. Such inspection is required to be made before the expiry of 14 days after the end of the Period of Maintenance.

The remedial or reinstatement works must be due to the 'use of materials or workmanship not in accordance with the Contract or neglect or failure on the part of the Contractor to comply with any obligation expressed or implied'. It is obviously preferable that the obligations are expressed rather than implied – there can then be no doubt about what is required. This applies particularly to reinstatement of trial pits in highways or pasture where damage or injury may result in imperfection.

Sub-clause (4), whilst containing a sanction, allows the Contractor by arrangement with and payment to the Employer to continue the process of reinstatement over an extended period. Climatic or ground conditions or the conditions of the excavated soil at the time when the reinstatement should be made, may make it impossible to achieve optimum compaction. Consolidation of the mass of fill will therefore continue over an extended period. It may be unrealistic and expensive for the Employer to require the Contractor to execute such work over a period of months after he leaves the Site so allowance for this can be made clear in the Specification.

Sub-clause (5) is a particular example of the application of clause 49. The continuing reinstatement of highways is a specialist application of designated materials not usually within the expertise of a ground investigation contractor. It is therefore preferable for the Contractor or the Employer to contract with the Highway Authority to supervise the work of primary reinstatement and to maintain the road surface levels thereafter for as long as the Authority may require. The structural integrity of the load–bearing components of the road pavement, that is the sub–base and base, is the ultimate responsibility of the Highway Authority. It follows that it may be preferable for the Authority to require the Employer to execute reinstatement under direction, or alternatively to execute the work, and maintain it, itself. Similar arrangements are applicable to owners or tenants of agricultural land, where the Contractor is kept aware of and required to maintain the levels and productive properties of the land. Given such action, the Period of Maintenance would not be unrealistically extended and, to repeat, liability for potential damage and injury will be a vicarious liability of those who undertake the work.

Rectification of defects in Ancillary Works

Separate from the preceding reinstatement and maintenance clauses, but nevertheless conjoined to them, is clause 50, relating to Ancillary Works.

> The Contractor shall if required by the Engineer in writing carry out such searches tests or trials as may be necessary to determine the cause of any defect imperfection or fault in any Ancillary Works under the directions of the Engineer. Unless such defect imperfection or fault shall be one for which the Contractor is liable under the Contract the cost of the work carried out by the Contractor as aforesaid shall be borne by the Employer. But if such defect imperfection or fault in any Ancillary Works shall be one for which the Contractor is liable the cost of the work carried out as aforesaid shall be borne by the Contractor and he shall in such case repair rectify and make good such defect imperfection or fault at his own expense in accordance with Clause 49.

This is straightforward where the Ancillary Works are

designed, constructed or installed by the Contractor. Where the Ancillary Works have been provided by a sub-contractor, the responsibility for what is probably a specialist operation, will devolve upon him through the Contractor.

Reporting

The Appendix to the Form of Tender specifies a stated period of weeks, after submission, within which the draft Report and final Report are to be approved, and refers to clause 14(5) which is quoted below. There is no express requirement in the *Conditions* for the draft and final Reports to be submitted within a specific time, although the final act of the Investigation is usually the submission of the final Report and there is a time limit for the Investigation. The Appendix to the Form of Tender may have to be modified to allow, and the Specification to explain, the requirements regarding times for submission of interim and draft Reports. The relevant clauses of the *Conditions* in relation to the Reports are quoted here.

Clause 1(1)(*n*) states that '"Report" means the report to be prepared and submitted in accordance with the Contract;'

Clause 14(1) refers to the programme to be submitted by the Contractor to the Engineer, the times allowed for each stage and, in its ultimate sentence, refers to Reports:

The programme submitted by the Contractor pursuant to this sub-clause shall take into account the period or periods for completion of the Investigation or different Sections thereof and the periods required by the Engineer for approval of testing schedules and reports which are provided for in the Appendix to the Form of Tender.

Clause 14(5):

If the time taken by the Engineer in giving his approval to the Contractor's proposed testing schedule or the draft Report or the final Report shall exceed the appropriate Period for Approval shown in the Appendix to the Form of Tender and as a result the time for completion of the Investigation or any Section thereof shall be exceeded the Engineer shall grant an extension of time in accordance with Clause 44 in respect of the time taken by him in excess of the appropriate Period for Approval shown in the Appendix to the Form of Tender which Period for Approval shall be calculated from the time elapsed between despatch of the Contractor's proposed testing schedule or the draft Report or the final Report as appropriate and receipt by the Contractor of the Engineer's respective disapproval consent or approval.

Clause 15(3):

If in addition to the superintendence in accordance with sub-clauses (1) and (2) of this Clause the Contract shall require or the Engineer direct the Contractor to make available on the Site or elsewhere the services of suitably qualified persons for description of soils and rocks logging of trial pits execution of geological and geotechnical appraisals other technical and advisory services and the preparation of technical reports the extent and scope of the service required shall be specified in the Contract.

Notwithstanding these clauses everybody involved in the Contract must understand that the Report is the object of the Investigation, and is covered by the Form of Tender which, among other things, binds the Contractor to the following:

We undertake to complete the whole of the Investigation comprised in the Contract within the time stated in the Appendix hereto.

Exceptions to this are described on p.77 at the beginning of this chapter. It will be noted that the Report is defined under clause 1(1)(*n*) as 'the report to be prepared and submitted in accordance with the Contract'. It does not define the draft report, which could be presented in several forms, depending upon the object of the Investigation.

The most exacting kind of Report is one which the Contractor, or the Engineer acting for the Contractor or as Expert Witness, prepares for consideration by lawyers. The Report may be used in legal disputes and as the lawyer's responsibility may be to present his case tactically he may ask that certain opinions or facets should be emphasised, reduced, expanded or deleted entirely. However, the potential Expert Witness must not depart from the facts or from his genuine opinion in his final Report. The final Report will be studied by the other party to any legal proceedings and may, or may not, be elaborated during the procedural processes of any legal action. Where legal requirements are involved it may not be possible to give a stated period for the preparation of the Report and a caveat such as 'dependent upon the requirements of Legal Advisors' may need to be introduced.

The Engineer should modify the Appendix to the Form of Tender to suit the sequence and form of reports which the type of Investigation demands. This should have been set down and explained in the Specification. Upon approval of the final Report, assuming that heralds the end of the Investigation, the Engineer has to issue the Certificate of Completion under clause 48(1).

Manner of reporting

It is not within the scope of this work to detail the manner or style of reporting. This is adequately covered by BS 5930: 1981, Section 7, clauses 39 and 40. In preparing the Specification, the Engineer will need to invoke particular sub-clauses within clauses 39 and 40. Often in the past, Reports have been prepared which were unsuited to the object of the Investigation. They were inadequate, had too much technical jargon, lacked clarity, failed to provide supporting information or were positively erroneous. The principal requirement is that the Report should be readable and understandable by all who need to use it. It may be necessary to divide it into several sections, starting with a Summary, followed by a discussion of the findings, an opinion and recommendations when required, and ending with illustrative and supporting test data.

The Specification can lay down the form of the Reports and the manner in which they are to be presented.

Rejection of Reports

As discussed in Chapter 3 (p.46), all reports, whether they are Site logs, preliminary, intermediate, draft or final Reports, require the approval of the Engineer under sub-clause 14(5) within the time limits set down in the Appendix to the Form of Tender.

The reason for non-approval, or rejection of a report would be failure or potential failure, to comply with the express, or implied, requirements of the Contract. An experienced Engineer would not normally reject a report merely because it lacked polish. His reason, or reasons, for rejection could include the factors listed below. In this illustration it is assumed that the type, sequence and content of the Reports have been detailed in the Specification. The Report on a small Investigation could be presented in letter form, but its contents would, nevertheless, be in the sequence given on p.46 and approval could be withheld for the same reasons.

Reasons for rejection of Site logs:

(a)   indecipherable writing;
(b)   written detail obscured by mud or washed out;
(c)   inadequate description of soils encountered;
(d)   no indication of boring progress by stated times;
(e)   sample numbers unrecorded;
(f)   levels of groundwater unrecorded;
(g)   no indication of use of casing;
(h)   no detail of *in situ* tests;
(i)   log unsigned.

Reasons for rejection of a preliminary Report:

(a)   inadequate information about the findings in relation to the object;
(b)   unsupported by plot plan;
(c)   unsupported by duplicates of Site Logs;
(d)   unsupported by preliminary boring logs;
(e)   no indication of the trend of the Investigation.

Reasons for rejection of a draft Report

(a)   fails to meet the object;
(b)   sequence and form of presentation incorrect;
(c)   unsupported by, or poor standard of, drawings, final logs and test data;
(d)   lacks clarity in discussion of factors;
(e)   illogical conclusions.

The Engineer will study the draft Report knowing that it is the tentative presentation by the Contractor of the content and form of the final Report. His comments on the draft Report should advise on the content and presentation of the final Report. He would not normally, for example, approve the draft Report and reject the final Report on points which were approved in the draft. To do so would invite claims by the Contractor for abortive work in preparation of the final Report.

Reasons for rejection of a final Report

(a) conclusions and recommendations very different from those presented in the draft Report;

(b) unacceptable standard of presentation.

It must be emphasised that the Contract *Conditions* lay down no criteria for the form, content or requirements of the Reports: these must be defined in the Specification.

Chapter 7
Payment and claims

Better it is that thou shouldest not vow than that
thou shouldest vow and not pay.
Ecclesiastes

This chapter deals with three topics: payment for contract work and variations; Employer's claims; and Contractor's claims. The variety of claims is infinite so only the most common types of claim will be considered.

Payment (machinery)

The Contract is by its terms (clause 56) a measure and value contract; unless substantial amendments are made, it is not a lump sum contract, although elements of the Investigation may need to be priced by individual sums. This creates a potential problem for the Employer: for him, the most valuable part of the Investigation will often be the Report but, for the Contractor, the most profitable parts of the Investigation will usually be those before the Report is prepared. There may, therefore, be little incentive other than the threat of the imposition of liquidated damages to encourage a speedy submission of the Report. One solution would be to ensure, by negotiation at tender stage, that a reasonably high figure is put against the item in the Bills of Quantity for submission of the Report.

It is principally the Site Operations part of the Investigation which will be susceptible to 'admeasurement'. As in the *ICE Fifth Edition*, any increase or decrease in quantities must be reflected in the sums to be certified and paid to the Contractor. There may be a tendency, particularly under the *Conditions* and in the context of clause 13(4), for tendering contractors to overprice items such as trial-boring. Excessive overpricing of individual items should be avoided, principally in the interests of the Employer but also in those of the Contractor. If the increase, or decrease, of the quantities in respect of any item renders any rates or prices unreasonable or inapplicable, appropriate adjustments should be made (clause 56(2)). This provision gives the Contractor some protection against the effects of a reduction in the quantities of items which he has overpriced in his tender: such a decrease in quantities might render the corresponding underpriced rates inapplicable. The corollary would be that if there were a large increase in quantities of the overpriced items, the rates or prices might then be inapplicable and be reduced accordingly. Measurements should be made in accordance with the current CESMM (clause 57).

The Bills of Quantity (unless stated otherwise) are 'deemed to have been prepared . . . according to the procedure set forth in the (current) "Civil Engineering Standard Methods of Measurement"' (clause 57). If, in error, items in the Bills are incorrectly described or omitted, the errors must be corrected by the Engineer and the resultant correction dealt with (and paid) as a variation (clause 55(2)).

Provision is made for joint consultation upon and attendance for measurements (clause 56(3)). The Contractor can be required to produce any back-up information to assist the Engineer in the carrying out of measurements. This could include driller's notes or logs.

The actual mechanics of interim payment are considered in Chapter 3, p.49–p.51. Sums due in respect of Nominated Sub-contractors must also be certified but should be shown separately in the certificate.

Any retention against interim payments must be agreed at contract stage by being stipulated in the Appendix to the Form of Tender (clause 60(4)).

Clause 60(5) provides for the release of the first half of the retention within 14 days of the completion of the Site Operations and for the second 14 days after the expiration of the Period of Maintenance. Because of these provisions there may be a period in which the Employer does not have the safeguard of retention. This arises because the Period of Maintenance expires within a period (usually six months) from completion of the Site Operations but much of the Laboratory Testing and the submission of Reports may not be completed until after the expiry of the Maintenance Period. This problem can be avoided by specifying a longer Period of Maintenance. Clause 60(5) makes provision for the release of retentions to correspond to the completion of parts or Sections of the Site Operations.

Clause 60(3) provides that the final Certificate of Completion is issued within three months of the submission by the Contractor of his final account which, in turn, must be submitted within three months of the Acceptance Certificate. The Acceptance Certificate is explained in clauses 61(1) and 61(2):

(1) Upon the expiration of the Period of Maintenance or where there is more than one such period upon the expiration of the latest period and when all outstanding work referred to under Clause 48 and all work of repair amendment reconstruction rectification and making good of defects imperfections shrinkages damage subsidence of backfill and other faults referred to under Clauses 49 and 50 shall have been completed and when Completion Certificates shall have been issued in respect of all Laboratory Testing and all Reports (if any) included in the Investigation the Engineer shall issue to the Employer (with a copy to the Contractor) an Acceptance Certificate stating the date on which the Contractor shall have completed his obligations to carry out the Investigation to the Engineer's satisfaction. (2) The issue of the Acceptance Certificate shall not be taken as relieving either the Contractor or the Employer from any liability the one towards the other arising out of or in any way connected with the performance of their respective obligations under the Contract.

No certificate in this Contract is final and conclusive, in the sense that all certificates are open to review and revision in arbitration. The value of any certificate can be corrected by the Engineer for good reason (clause 60(7)). Failure to certify, or to pay properly and on time, gives the contractual right to interest to the extent of late or under-payment (clause 60(6)).

Payment (variations)
The variation clauses are similar to those in the *ICE Fifth Edition*. The Engineer should order any variation to any part of the Investigation which is necessary or desirable for the completion of the Investigation. The ambit of clause 51(1) is wide and covers such diverse alterations as the need for drilling techniques different to those specified, or reports which are required to be submitted in a format different to that originally specified, altering boring positions or ordering acceleration of part of the Investigation.

The principal limitation on the variation clause is that no variation can be ordered unless it is necessary or desirable for the completion of the Investigation (as stated in clause 1(6) and in other contractual documents). Thus, if the Investigation is specified to relate to site A, the Engineer cannot usually insist by variation order that the Contractor works on site B; or, if the Investigation involves auger-boring and a Report only, the Engineer cannot require the Contractor to build a bungalow on the Site.

The valuation of variations is governed by clause 52. Valuations and certifications must take account of amounts properly due in respect of variations.

Variations, subject to one important (and new) exception in this *Contract*, must be ordered by the Engineer. Although there are requirements that such orders are in writing, clause 51(2) envisages and allows valid orders to be given orally. The exception is the new clause 13(4), complemented by clauses 13(5) and (6) which state:

(4) If during the execution of Site Operations the Contractor shall encounter ground or geological conditions which in his opinion make it necessary for the effectiveness of the Investigation or for the adequacy of the Report to continue the operations of boring drilling excavation sampling or in situ testing to a greater depth than is included in the Schedules before Equipment is moved from the position of the borehole drill hole or excavation and the Engineer's or (subject to the limitations referred to in Clause 2) the Engineer's Representative's instructions cannot be immediately obtained the Contractor may continue such operations or change the mode of operation at his own discretion provided always that the cost of such continued operations or changed mode of operation shall not exceed such sum as may have been agreed between the Engineer and the Contractor in writing at the commencement of the Site Operations unless a further instruction or specification shall have been subsequently issued by the Engineer in accordance with sub-clause (5) of the Clause.
(5) In the event of operations being continued in accordance with sub-clause (4) of this Clause every endeavour shall simultaneously be made to obtain the Engineer's or (subject to the limitations referred to in Clause 2) the Engineer's Representative's instructions.
(6) Subject to any instruction of the Engineer previously given pursuant to this sub-clause such continued or changed operations shall be deemed to have been carried out as a variation ordered pursuant to Clause 51, provided always that if the Engineer shall decide that the Contractor was not justified in continuing or changing the operation in accordance with sub-clause (4) of this Clause the Engineer shall so inform the Contractor in writing and shall issue further instructions or specification as he may think fit to govern how any similar ground or geological conditions which may be encountered in the future shall be dealt with.

Although these clauses introduce the novel concept of non-instructed variations, they can be operated successfully given an honest and responsible contractor. The criteria for proceeding justifiably to the additional Site work without instruction are, firstly, that ground geological conditions make it necessary for the effectiveness of the Investigation or for the adequacy of the Report and, secondly, that the Engineer's or his delegate's instructions cannot be obtained immediately. The Contractor takes the slight risk that the results of his additional work or changed method of operation may show that the alterations were unnecessary.

Some difficulties for the Employer (and corresponding benefits for the Contractor) arise from these sub-clauses. An additional foot or two's depth of boring or the odd extra *in situ* test can usually be justified as necessary for the adequacy of the Report. The onus of proving that the continued or changed operation was not necessary seems to rest on the Engineer (and thus on the Employer). This may not be easy because, unavoidably, the Engineer was not present at the relevant time. The Employer's best protection against abuses is to ensure (probably by the original specification) that an appropriate financial limit is imposed on any expenditure otherwise allowable under these sub-clauses. The obvious advantage to the Employer is that it reduces the risk of claims for expensive waiting time pending the instructions of the Engineer.

Employer's claims (liquidated damages)

Clause 47 provides for the imposition of liquidated damages in the event of, and to the extent of, delay on the part of the Contractor in completing the whole or Sections of the Investigation. The clause is, however, grammatically and arithmetically complicated.

The object of the liquidated damages clause is to enable the parties to ascertain at contract stage what sum, at a daily or weekly rate, will represent the damages due to the Employer in respect of the Contractor's failure to complete the Investigation within the prescribed or any extended time (clause 47(1)(*b*)).

Clause 47(1)(*a*) states:

In the Appendix to the Form of Tender under the heading 'Liquidated Damages for Delay' there is stated in column I the sum which represents the

91

Employer's genuine pre-estimate (expressed as a rate per week or per day as the case may be) of the damages likely to be suffered by him in the event that the whole of the Investigation shall not be completed within the time prescribed by Clause 43.

Provided that in lieu of such sum there may be stated such lesser sum as represents the limit of the Contractor's liability for damages for failure to complete the whole of the Investigation within the time for completion therefor or any extension thereof granted under Clause 44.

The late completion of a ground investigation will often have serious financial consequences. It would be legitimate to include the following types of loss in the genuine pre-estimate for liquidated damages.

(*a*) Cost of extended involvement of Employer's and Engineer's staff or specialists retained by the Employer, over any period of delay (even where a local or government authority is the Employer).

(*b*) Extra licence fees or rental for excess time on the Site.

(*c*) If the contract is on a fluctuation basis, allowances for increases in cost of labour, materials, plant and equipment over any period of delay.

(*d*) Foreseeable additional financing charges attributable to delay in respect of the proposed development whose potential depends upon the outcome of the Investigation.

(*e*) Foreseeable loss of profit in the proposed development to be occasioned by delay.

(*f*) Additional costs of the proposed works which would foreseeably be caused by delay (for instance, delay in the Investigation might well mean that the groundworks for the proposed development had to be carried out in winter).

Whilst allowance is made for the parties to limit the rate for liquidated damages to a figure which is less than a genuine pre-estimate of the likely delay-related loss, the obverse is not true. If a rate of liquidated damages is imposed which is obviously and substantially in excess of what a genuine pre-estimate could have been, it will be considered a 'penalty' in law which is void and unenforceable (Dunlop v Selfridge & Co 1915 AC 847)

Where completion in Sections has been agreed, clause 47(2) provides the appropriate formula and machinery for imposing liquidated damages for late completion of particular Sections. In a ground investigation contract, this allows the necessary flexibility to protect the Employer. For instance, any delay in the Site Operations, or of a particular part thereof, will result in particular loss (e.g. extended rental of the Site); the pre-estimate for that Section of the Investigation can allow for that loss; (clause 48(2)(i)(*a*) dictates that the whole of the Site Operations are considered to be a Section; the Appendix to the Form of Tender leaves the appropriate gaps to be filled in by the parties). In many cases, delay in delivery of the Report will lead to serious financial losses; provided that the net amount allocated in liquidated damages in respect of late delivery of the Report is a genuine pre-estimate, the apparent disproportion will be acceptable at contract stage.

Clauses 47(1)(*b*), 47(2)(*b*)(ii) and 48 allow for the early completion of parts of the Investigation and a proportionate reduction of the rates of liquidated damages for delays to the remaining elements of the Investigation.

By clause 47(3), liquidated damages are not payable or allowable to the Employer until and unless:

(*a*) The original overall time or the time for completion of a Section or any previously extended times, have passed.

(*b*) The Engineer has assessed under clauses 44(3) or 44(4) (on his own initiative) whether any and, if so, what extensions of time are due.

(*c*) The Engineer has notified the parties of his opinion that the Contractor is entitled to, or is not entitled to, a further extension of time.

If, however, further extensions are granted after liquidated damages have been paid or allowed to the Employer, appropriate adjustments and repayments are provided for in clause 47(5).

Employer's claims (defects)

The Contractor's obligations with regard to the quality and extent of the Investigation and the Employer's rights to enforce those obligations during the Contract are considered in Chapters 4 and 6. This section deals with defects in the Investigation which have not been put right by the Contractor during the Investigation.

If, during the progress of the Site Operations or during the

93

Period of Maintenance, the Contractor fails, or refuses, to comply with a proper order from the Engineer, under clauses 39(1) or 49(2), to remove, repair or replace defective materials or work, the Employer may employ someone else to effect the order. He is entitled by clauses 39(2) and 49(4) to claim (or deduct from sums otherwise due to the Contractor) any loss and expense which he incurs as a result.

If defective work or materials (i.e. not in compliance with the Contract) have not been rectified or discovered before the expiry of 14 days from the end of the Period of Maintenance, the Employer has no express contractual obligation to give the Contractor the opportunity to rectify it. However, at common law, the Employer is bound to mitigate the consequences of the Contractor's breach of contract; although this duty is not onerous, the Employer should usually give the original Contractor the opportunity to put right the defect at his expense unless there is some good reason why he should not do so. A good reason might be that the remedial works had to be done as an emergency, or that the Contractor had already been given a chance to make the repair. If the Employer does employ others to do the remedial work, assuming no failure to mitigate, he is entitled to the costs of, and losses occasioned by, the remedial works.

Any breach of contract on the part of the Contractor which causes loss or damage to the Employer which, at the date of the Contract, was a foreseeable or, actually foreseen, consequence of such a breach, will entitle the Employer to claim such loss or damage as damages from the Contractor (Hadley v Baxendale 1854 9 EX 341).

If the Report has been prepared in breach of the requirements of the Specification or with lack of care, it may be many years before the consquences are felt by the Employer. The effect of statutory limitation periods on such claims is considered in Chapters 1 (p.10) and Chapter 9 (p.117). Subject to limitation, provided the types of loss suffered by the Employer were actually foreseen, or were reasonably foreseeable, they are recoverable despite the fact that they may be incurred years later. Thus, a careless and inaccurate recommendation in the Report that strip foundations are appropriate for an old rubbish fill site could well make a Contractor liable for any of the following losses:

(*a*) costs of suspending the work pending a re-design of the foundation;
(*b*) extra professional costs for re-designing the foundations;
(*d*) wasted expenditure on work done before suspension;
(*e*) loss of profit or rental on a commercial or local authority building development.

Contractor's claims

This section considers the principal types of claim which a ground investigation Contractor is likely to make. The fact that there are many clauses which give rise to contractual claims (which are in effect claims for contractual debts) does not exclude the Contractor's rights at common law to claim damages for breach of contract. Some examples of such claims are given on p. 94. Most of the contractual claims clauses entitle the Contractor to 'cost' which is defined by clause 1(5):

> The word 'cost' when used in the Conditions of Contract shall be deemed to include overhead costs whether on or off the Site except when the contrary is expressly stated.

Accordingly, save where stated otherwise (e.g. clause 12(3)), there is no contractual entitlement to loss of profit.

Clause 7 claim

The designation or establishment by the Engineer of bench marks and on-site reference points and the timely provision by the Engineer of information and instructions have been considered in detail in Chapter 3 (p. 40). Failure or inability to comply with such obligations gives rise to valid claims under clause 7(4):

> If by reason of any failure or inability of the Engineer to issue at a time reasonable in all the circumstances drawings schedules or instructions requested by the Contractor and considered necessary by the Engineer in accordance with sub-clause (1) of this Clause or failure by the Engineer to establish bench-marks or on-site reference points the Contractor suffers delay or incurs cost then the Engineer shall take such delay into account in determining any extension of time to which the Contractor is entitled under Clause 44 and the Contractor shall subject to Clause 52(4) be paid in accordance with Clause 60 the amount of any such cost as may be reasonable. If such drawings schedules or instructions require any variation to any part of the Investigation the same shall be deemed to have been issued pursuant to Clause 51.

95

The formula used in this sub-clause is similar to other claim clauses.

It entitles the Contractor to:

(*a*) reasonable delay-related cost;
(*b*) reasonable non-delay-related cost;
(*c*) an appropriate extension if there has been, or will be, delay.

Obviously the delay and the cost must be caused by, or properly attributable to, the original failure or inability of the Engineer. Not all cost is necessarily reasonable.

The Contractor is also obliged to follow the claims procedure in clause 52(4) of which the relevant sub-provisions are:

(*b*) If the Contractor intends to claim any additional payment pursuant to any Clause of these Conditions other than sub-clauses (1) and (2) of this Clause he shall give notice in writing of his intention to the Engineer as soon as reasonably possible after the happening of the events giving rise to the claim. Upon the happening of such events the Contractor shall keep such contemporary records as may reasonably be necessary to support any claim he may subsequently wish to make.

(*c*) Without necessarily admitting the Employer's liability the Engineer may upon receipt of a notice under this Clause instruct the Contractor to keep such contemporary records or further contemporary records as the case may be as are reasonable and may be material to the claim of which notice has been given and the Contractor shall keep such records. The Contractor shall permit the Engineer to inspect all records kept pursuant to this Clause and shall supply him with copies thereof as and when the Engineer shall so instruct.

(*d*) After the giving of a notice to the Engineer under this Clause the Contractor shall as soon as is reasonable in all the circumstances send to the Engineer a first interim account giving full and detailed particulars of the amount claimed to that date and of the grounds upon which the claim is based. Thereafter at such intervals as the Engineer may reasonably require the Contractor shall send to the Engineer further up to date accounts giving the accumulated total of the claim and any further grounds upon which it is based.

(*e*) If the Contractor fails to comply with any of the provisions of this Clause in respect of any claim which he shall seek to make then the Contractor shall be entitled to payment in respect thereof only to the extent that the Engineer has not been prevented from or substantially prejudiced by such failure in investigating the said claim.

Clause 12 claim
Clause 12 is traditionally, in the *ICE Fifth Edition*, regarded as

'the Contractor's friend'. The wording is not significantly different to the clause 12 in the *Conditions*, sub-clause 1 of which is as follows:

If during execution of the Site Operations the Contractor shall encounter physical conditions (other than weather conditions or conditions due to weather conditions) or artificial obstructions which conditions or obstructions he considers could not reasonably have been foreseen by an experienced contractor and the Contractor is of opinion that additional cost will be incurred which would not have been incurred if the physical conditions or artificial obstructions had not been encountered he shall if he intends to make any claim for additional payment give notice to the Engineer pursuant to Clause 52(4) and shall specify in such notice the physical conditions and/or artificial obstructions encountered and with the notice if practicable or as soon as possible thereafter give details of the anticipated effects thereof the measures he is taking or proposing to take and the extent of the anticipated delay in or interference with the execution of the Investigation.

The Clause relates only to the Site Operations and, therefore, can relate only to operations 'on under in or through the Site' (clause 1(1)(*k*)). It can be applied, however, to conditions and obstructions encountered during the Period of Maintenance.

Although the words are not very different from the *ICE Fifth Edition*, the general application may be different, depending upon the type of ground investigation. For example, in a case where an experienced contractor could reasonably have expected that there would be no discernible sequence of strata (e.g. periglacial, landslip alluvial or deposits of similar origin); the fact that during trial-boring he discovers more boulders or saturated sand than he anticipated will not entitle the Contractor to the benefit of clause 12. Any clause 12 claim will be looked at in the context of the following points.

(*a*) The information made available at tender stage;
(*b*) The information discovered as a result of what the Contractor is deemed to have done under clause 11(1) (e.g. inspection and examination of the site and surroundings, general consideration of known geology – see Chapter 2, p.19);
(*c*) The extent, object and type of Investigation required under the Contract (e.g. where a site is known to be clay and the object of the Investigation is to determine its

97

classification and engineering properties to a given depth, running sand may, in that context, be an unforeseeable physical condition);

(d) General geotechnical knowledge (e.g. if it is known that there is limestone which is subject to ground water-flow, it would be reasonably foreseeable that some of the limestone would be weathered).

There is, in clause 12(2), a variety of courses open to the Engineer once he has been notified by the Contractor of the offending physical conditions or artificial obstructions. The most important will be instructions on how to deal with the problems and instructions for variations.

Clause 12(3) follows the usual contractual claim formula with entitlements to extensions of time and financial reimbursement. Two points are worthy of comment:

(a) The Contractor is entitled to a 'reasonable percentage addition' on the recoverable cost in respect of profit;
(b) Irrespective of whether the Engineer exercised any of the options allowed in clause 12(2), the Contractor is entitled to the reasonable cost of overcoming the encountered problems, and also the delay and disruption cost.

Clause 13(3) claim
The origins of the clause 13(3) claim are first found in clause 5:

(5) The several documents forming the Contract are to be taken as mutually explanatory of one another and in case of ambiguities or discrepancies the same shall be explained and adjusted by the Engineer who shall thereupon issue to the Contractor appropriate instructions in writing which shall be regarded as instructions issued in accordance with Clause 13.

This is identical to the *ICE Fifth Edition.*
Clause 13(1) states:

Save in so far as it is legally or physically impossible the Contractor shall carry out the Investigation in strict accordance with the Contract to the satisfaction of the Engineer and shall comply with and adhere strictly to the Engineer's instructions and directions on any matter connected therewith (whether mentioned in the Contract or not). The Contractor shall take instructions and directions only from the Engineer or (subject to the limitations referred to in Clause 2) from the Engineer's Representative.

What is 'legally impossible' is easily understood. It relates to the situation which occurs when an Investigation can only be carried out in accordance with the Contract by contravening statute or common law. For instance, if the Specification required the dumping of contaminated extrusions from boring onto land belonging to a person who had not given permission for this, the dumping would involve trespass at common law; compliance with that part of the Contract would be legally impossible. If the design of Ancillary Works did not comply with building regulations (if applicable), the Contractor could not both perform the Contract and comply with the law.

'Physically impossible' might appear to be an absolute term but very little in civil engineering or ground investigation is absolutely impossible, given time, money and patience. In a recent unreported decision (Turriff v City of Chester, unreported 1978), the judge (his Honour Judge W. Stabb QC, official referee) held that these words had to be considered as meaning effectively impossible in the context of the contract; they did not necessarily mean impossible in the absolute sense. Some examples of what would be physically impossible are given here.

(a) If a light cable percussion rig were specified and encountered solid granite, continuance with that which the Contract required would not be physically possible.

(b) If, on a sloping site, piezometers were specified to be installed at certain positions and certain levels and, before or during the installation, the land slips, then that installation, whilst still physically possible at different positions or at different levels, cannot physically be carried out in accordance with the Contract.

Once the physical or legal impossibility has been ascertained, the Engineer must issue instructions to overcome it. Failure to issue instructions reasonably, promptly, or at all, will lead to a valid claim by the Contractor against the Employer under clause 7(3) for damages. The consequences of the issue of instructions is contained in clause 13(3):

If in pursuance of Clause 5 or sub-clause (1) of this Clause the Engineer shall issue instructions or directions which involve the Contractor in delay

99

or disrupt his arrangements or methods of working so as to cause him to incur cost beyond that reasonably to have been foreseen by an experienced contractor at the time of tender then the Engineer shall take such delay into account in determining any extension of time to which the Contractor is entitled under Clause 44 and the Contractor shall subject to Clause 52(4) be paid in accordance with Clause 60 the amount of such cost as may be reasonable. If such instructions or directions require any variation to any part of the Investigation the same shall be deemed to have been given pursuant to Clause 51.

Clause 14(5) claim (late approvals)

The *Conditions* provide, in the Appendix to the Form of Tender, time limits for the approval by the Engineer of the Contractor's testing schedules, draft Report and final Report. Clause 14(5) entitles the Contractor to an extension if these time limits are exceeded, but it does not expressly entitle him to financial reimbursement. If there is no good reason for the delay in approval, the Contractor would be entitled to damages from the Employer for the failure to the Engineer to grant the approvals within the agreed times.

Clause 42(1) claim (late possession)

Chapter 3 (p.47) amplifies the Employer's obligation to grant the Contractor possession of parts of the Site on time. Clause 42(1) gives the Contractor the right to claim:

If the Contractor suffers delay or incurs cost from failure on the part of the Employer to give possession in accordance with the terms of this Clause then the Engineer shall take such delay into account in determining any extension of time to which the Contractor is entitled under Clause 44 and the Contractor shall subject to Clause 52(4) be paid in accordance with Clause 60 the amount of such cost as may be reasonable.

This clause does not exclude the Contractor's common law rights; the Contractor can also proceed against the Employer for damages (as opposed to contractual reimbursement) for delayed or hindered possession.

Non-contractual claims

Such claims are infinite in variety so full consideration of every type of claim is not possible. Examples of common claims are given below.

(a) An employee or agent of the Employer (other than the Engineer or Engineer's Representative) may ask the Contractor to carry out additional work. The claim does not arise because within the contract only the Engineer may issue variations. However, the claim is recoverable from the Employer.

(b) When the Employer has positively failed to provide reasonable co-operation, the Contractor may claim losses which result from that lack of co-operation as damages for breach of an implied term of the Contract that the Employer should provide reasonable co-operation.

(c) Where the Engineer has refused to issue instructions which he was obliged by the Contract to issue, the claim by the Contractor is for damages for breach of contract by the Employer.

(d) Where the Employer has wrongly interfered with the issue and content of certificates, a claim for breach of contract (of the implied terms of co-operation) can be sustained.

Forfeiture and determination

Under the forfeiture clause (clause 63), the Employer has the right to expel the Contractor from the Site and to complete the Investigation himself. This can only be done if one of the specified grounds for forfeiture exists (e.g. bankruptcy, receivership, abandonment, continued dilatoriness after warning by Engineer) and the procedure laid down is complied with. Provision is made for financial adjustments to ensure that the Engineer is paid or allowed all additional costs of completing the Investigation (clause 63(4)).

At common law, if either party repudiates the Contract and the innocent party accepts the repudiation, the Contract is effectively determined. Repudiation arises where one party indicates an intention no longer to be bound by an important term of the Contract (White and Carter (Councils) Ltd v McGregor 1962 AC 413). In the ground investigation Contract, examples of repudiatory conduct might be one of the following:

(a) The Contractor prematurely (and without good cause) removes all his equipment from the Site.

(b) The Employer, through the Engineer, refuses (without

101

good cause) to designate or establish bench marks and on-site reference points to enable the Contractor to set-out and start the Site Operations.

(*c*) The Employer persistently fails to honour interim certificates.

Chapter 8
Codes of practice – their relevance and application

Multiplication is vexation
Division is as bad
The Rule of Three doth puzzle me
And Practice drives me mad.
                    Anon

British Standards
The British Standards Institution in their publication BSO: Part 1: 1981 gives a full explanation of 'general principles of standardisation in terms of the aims of British Standards and the limitations and legal application of their use'. Of course, this covers all standards from shirt buttons to concepts of structural design.

Of particular importance to the *Conditions* for ground investigation is the unequivocal statement in paragraph 7.1 that:

British Standards are publicly available documents voluntarily agreed as a result of processes of public consultation designed to secure public acceptance.

However, the publication of a standard by the BSI does not ensure its use. Its application depends on the voluntary action of interested parties. It becomes binding only if a claim of compliance is made, if it is invoked in a contract or if it is called up in legislation. In this case, the operative word in a contract is 'invoked'.

BSO: Part 1: 1981, in paragraph 7.3 under 'Duty of Care', declares:

The care exercised in the production of standards is relied upon by the users of the standard who themselves owe a similar duty to the public. It remains the responsibility of users to ensure that a particular standard is appropriate to their needs. Within their scope, national standards provide evidence of an agreed 'state of the art' and may be taken into account by the Courts in determining whether or not someone was negligent.

It does not necessarily follow, however, that non-compliance with a code of practice automatically implies negligence.

Contractual use of Standards

Section 8 of BSO: Part 1: 1981 covers the application of standards to contracts and the following extracts are, again, of particular relevance. Paragraph 8.1 includes this sentence:

> Standards or parts of standards invoked in contracts become legally binding on the contracting parties. However, no British Standard can purport to include all the necessary provisions of a contract.

In ground investigation, the reverse of the second sentence may apply, and care must be taken to ensure that ambiguity in the specified method of operation or test does not apply. This is foreseen by the third paragraph of Section 8 which states that:

> Many British Standards contain options and other matters that need to be clarified by additional contractual provisions, when invoked in contract specifications.

If a particular code of practice is incorporated without qualification into the contract, and that code contains optional alternatives (e.g. compaction testing in BS 1377: 1975), there will be confusion. Further, if the Engineer insists that the Contractor complies with, or follows, one of the alternatives that will probably be considered to be a variation and if the Contractor can demonstrate a financial disadvantage as a result of following the alternative, as instructed, he can properly mount a claim under clause 52 or clause 13(3) (Neodex v Borough of Swinton and Pendlebury 5 BLR 34, Crosby v Portland UDC 5 BLR 121).

Lastly, and most importantly, the tendency of draftsmen of contracts to generalise with reference to a standard (or standards) is warned against by paragraph 8.4, which states:

> Some British Standards are deliberately drafted in advisory form, i.e. codes of practice, guides and recommendation, and are therefore inappropriate for simple reference in contracts.

Codes applicable to ground investigation

Unfortunately, ground investigation involves the application of several codes of practice and British Standards which relate to guidance, method and performance of an investigation, and also to specific requirements in the execution of *in situ* and Laboratory Tests and samples obtained from the Investigation.

These can be divided into two categories which will be discussed on p.106–114.

Categories of British Standard Codes of Practice and British Standards

As previously stated, codes of practice are presented as guidance in standards of good practice. This is now made clear in individual codes by a statement of the standing of the code. For example, that in BS 5930: 1981 (*Code of practice on site investigation*) states:

> This code of practice represents a standard of good practice and therefore takes the form of recommendations. Compliance with it does not confer immunity from relevant statutory and legal requirements.

It is unfortunate that this disclaimer is usually printed inside the cover page of the publication. It would be more effective on the title page, and underlined.

It may be accepted that the *Conditions* envisage, by implication, that, unless otherwise indicated, the work is to be undertaken in accordance with the recommendations of the relevant code(s) of practice as far as quality is concerned. Reading the codes with the *Conditions* can lead to confusion and ambiguity in interpretation. To avoid this, it is essential that the Specification and Bill of Quantities, when detailing options and tests, specify that the operation or test shall be executed in accordance with the recommendations of a particular British Standard code of practice, and that the relevant clause, sub-clause, table or figure is quoted. There should be thus no ambiguity, but it should also be understood that, where the ground conditions do not permit the operation to be executed exactly in accordance with the particular part of the specified code, any variation from it will have to be covered by a variation under clause 51(1) or an instruction under clause 13 of the *Conditions*. Given this, there is a specific requirement in performance to achieve the objectives of the investigation and, if varying methods or tests are employed arises, a basis on which to discuss any dispute which may arise during or after the completion of the ground investigation.

Codes of practice directly relevant to ground investigation are listed below in the order in which they apply to the specification and to the execution of the site works:

Category 1
BS 5930: 1981 *Code of practice on site investigation*
  (formerly CP 2001).
CP 2004: 1972 *Foundations*
BS 6031: 1981 *Code of practice for earthworks,*
  (formerly CP 2003: 1959).
BS 5573: 1978 *Code of practice for safety precautions in the construction of large diameter boreholes for piling and other purposes* (formerly CP 2011).

Currently in draft form under code EPC/47 is a *Code of practice for 'The identification and investigation of contaminated land'*. Whether this will be incorporated with BS 5930: 1981 or be published as a separate code is not yet known. In either case, it is unlikely to be published before this book.

Category 2
BS 812 (Part 1–4) *Methods for sampling and testing mineral aggregates, sands and fillers.*
BS 1377: 1975 *Methods of test for soil for civil engineering purposes.*
BS 1924: 1975 *Methods of test for stabilized soils.*

Application of codes in category 1
BS 5930: 1981 *Code of practice on site investigation*
This is the master code for use in support of the *Conditions*, and *vice versa*. An intimate knowledge of it is essential to the preparation of the Contract documents and the execution of the Contract. Published in 1981, it has replaced the Institution of Civil Engineers Code of Practice CP 2001: 1957 but it is possible that some investigations may be of a forensic nature, where a failed structure or land mass may have been designed from data obtained from an initial investigation based upon CP

2001. Reference to the previous code may therefore be necessary background to the planning and execution of the contract.

In most cases, users of a code of practice of this type refer immediately to the section which applies to the problem in hand. Using the document in this way may lead to misapplication so the reader is advised to study the opening sections which describe the sequence of investigation, and the general procedure which is described in the body of the Code. The introduction to BS 5930: 1981 deal mainly with the selection of construction sites. It has less to say about the wider concept of investigation which, in the preparation of county or district development studies by local authorities, government departments or their consultant advisors, may involve large stretches of countryside, the allocation of areas for mineral works, landscaping and development, with further sub-divisions into industrial and residential zones. These require an overall site investigation, and in this context the code gives a positive direction and should be used as a basis where the Contract requires site investigation. BS 5930: 1981 offers excellent guidance for the landscape planner, the town planner, the mineralogist and the architect, surveyor and estate agent preparing individual studies based on these *Conditions*.

A distinction is now made between the terms 'site investigation' and 'ground investigation'. 'Site investigation' covers not only the ground conditions particular to the Site but includes the immediate geological and geographical area within which the Site is contained. It therefore includes information on communications relating to the Site, its geology and its history. The geological study and ground exploration is called the 'ground investigation' and is dealt with in detail.

The code is explicit on the experience and qualifications required of personnel involved at each stage of the work, those responsible for: the several methods of exploration, sampling and testing; detailed classification and determination of the engineering properties of the soils encountered; the presentation of the Report. It also includes specialists in the fields of contamination due to toxic, radioactive or other agents which may be encountered in contaminated land. The code goes as far as it can to ensure that the Employer and Engineer are reassured that, at each stage, the right man has been applied to the right job. This is reinforced by clause 15(3) of the *Conditions* as discussed in Chapter 1, p.00.

The Code contains eight Sections, supported by eight Appendices. It is not intended as a text book, although several Sections, particularly those relating to operational methods and tests in boreholes, detail the requirements.

Sections 1 and 2 present general considerations and discuss the application of the art, review the conditions which may be encountered, and discusses the planning and primary objectives of site investigation.

The term 'desk study' is used for the first time. It describes the preliminary studies made to assess the geology, topography and communications of the area. From this study, together with Site reconnaissance, a discussion in the form of a preliminary Report can be presented. The preliminary Report will give an assessment of the situation and, when necessary, making specific recommendations for further investigation. Recommendations could include the investigation of salient features within the area. The Report might recommend a patterned investigation or a sequence of regularly spaced investigation points, in, for example, the choice of road alignment and in the zoning of areas for particular use.

General considerations and procedure are covered by Sections 3 to 6 and form the main body of BS 5930: 1981. These sections deal with 'ground investigation', including 'excavation, boring, sampling, probing and tests in boreholes', 'Field tests,' and 'Tests on samples'. Graphic illustrations of the construction and arrangement of apparatus to be employed in testing, together with formulae, calculations and related graphs are included. They offer guidance to the Engineer checking laboratories, as required by clause 8(3) of the *Conditions* (discussed in Chapter 2), and for technicians engaged in Site Operations and *in situ* testing.

Sections 4 to 6 deal with the preliminaries to the construction of civil engineering and building work. They describe a logical approach to any problem involving the use of land in such a way that the Engineer can develop an approach appropriate to his own requirement.

Reports and interpretation are covered by Section 7, which deals with the presentation of the results of work (the subject of

Chapter 6 of this book). It should be studied by those responsible for presenting the results of an investigation to any Employer, from the lay person to the geotechnical engineer, and also by technicians responsible for producing field reports and carrying out site and laboratory testing. Section 7 includes recommendations for types of site log, daily reports and data sheets for *in situ* tests. Methods of presentation of the detail revealed by trial pits, light cable percussion, boring, rotary core drilling and examination of exposures are illustrated by the inclusion of recommended forms of log, including operative logs. There are also examples of description and classification terms, annotation and the use of graphic symbols. If this code is incorporated in the Contract, the Employer can expect a report which presents details on the code forms; if different forms or a different classification are required, these must be specified elsewhere.

Section 8 is devoted to the description of soils and rocks. The methods of soil description and classification are now based upon the *British Soil Classification System for Engineering Purposes* (BSCS), developed by the Road Research Laboratory. This has been modified to give field identification and description of soils as an itemised list and to give laboratory identification information in detail. The coded presentation of soils, extended from the Casagrande Classification, is particularly useful. The soil and rock classifications are interdependent as are the graphic symbols used. Thus, in the presentation of information on boring logs, or borehole sections, both classifications may be used and will be complementary in interpretation.

In Figure 1, Chapter 1, 'Cost Indication' is given as a stage at which it may be decided to proceed with an investigation, or a section of it, or not. In this connection, it may be noted by those familiar with the previous code (CP 2001: 1957) that there is no specific indication of cost. This has, in the past, led to unfortunate situations where the estimated cost had been taken as one per cent of the total value of the works. Now, however, the cost of investigation is related to the extent of the proposed construction and to the complexity of the ground. It demonstrates that no sensible indication of the cost of a full investigation can be made until completion of the desk study.

CP 2004: 1972 *Foundations (currently under review);*
BS 6031: 1981 *Code of practice for earthworks and that currently in draft relating to the investigation of contaminated land (EPC/47)*
These are considered to be 'indicative' in that the type and extent of the works which may follow are dependent upon the result of the investigation and indicate how far the investigation should be taken. These codes overlap and complement BS 5930: 1981. They lead the Engineer from consideration of his construction, and the design criteria involved, to the specific data required from a ground investigation and the methods necessary to achieve them. At this stage the Engineer or architect responsible for construction may delegate the responsibility for the Site and/or ground Investigation to the geotechnical engineer or engineering geologist. It may be relevant for the Contract to refer to these codes in the context of the recommendations to be made by the Contractor in the Report.

If the Engineer or architect responsible for construction attempts to cover all contingencies by invoking more than one code for one investigatory operation, ambiguity is almost certain to occur, with consequent misunderstanding and confusion. So, although reference to that investigatory operation is made in the construction code, the reference invoked should be from BS 5930: 1981 and clause, sub-clause, table or figure number should be given. For example, if BS 6031: 1981 were to be used as an operational directive such as 'generally in accordance with . . . etc.', the Contractor could be given two options, since in Sections 2 and 3, the code deals separately, with 'Cuttings and embankments, grading and levelling' and 'Trenches, pits and shafts'. When dealing with site investigation Section 2 refers back to BS 5930: 1981, but describes the methods of investigation in detail. Section 3 refers back to Section 2. Thus, although BS 6031: 1981 gives adequate guidance on the investigatory methods necessary to obtain information, the particular operation to be invoked in the Contract should be that given in BS 5930: 1981.

CP 2004: 1972, site exploration and soil tests are discussed in Section 2 and reference is made to the then operative CP 2001: 1957. But again, the code is advisory, and it is necessary to refer back and invoke BS 5930: 1981 in the Contract.

BS 5573: 1978 Code of Practice for safety precautions in the construction of large diameter boreholes for piling and other purposes Formerly CP 2011, this code relates to clause 19(1) and (2) of the *Conditions* in the excavation of boreholes of, or exceeding, 0.7 metre (2 feet 4 inches approximately) in diameter. These are likely to be used where the borehole is to be incorporated in a later construction, or when the Engineer requires visual examination of the strata *in situ*. The latter could apply when it is necessary to determine the exact position of a slip plane or zone and to extract undisturbed soil samples from that point and the surrounding area. It would also assist in obtaining photographic evidence in support of a claim under clauses 11 and 12 of the *Conditions* should the matter proceed to litigation.

In the context of safety the term 'borehole' includes the excavation of 'trial pits' as discussed in BS C.P. 6031: 1981, Section 3. Before invoking requirements or limitations from one of these codes, the Engineer will have to decide which is applicable. If there is any doubt about how appropriate either code is, he should, with legal advice, draft a requirement related to safety with due regard to his responsibilities should he specify the method of construction.

Applicable in extension of this code are BS 1397: 1976 'Specification for industrial safety belts, harnesses and safety lanyards' and BS 2830: 1973 'Suspended safety chains and cradles for use in the construction industry'.

Contaminated land and made ground are dealt with in BS 5930: 1981 but not to the extent necessary now that more use is being made of sites which were previously regarded as being unsuitable or undevelopable. Recent ground investigations, particularly of made ground and tip sites, have encountered combustible and toxic materials which are a hazard in the investigating operations and to the public during the Investigation and, unless adequate provision is made in the design and planning of the Investigation, may affect the public after construction. It is re-emphasised that the Employer, the Engineer and the tenderers must be made aware of this. Occurrences consequential to, or arising from, the ground investigation operations, such as combustion due to ignition of the exposed material or gases from it, may cause delays and damage, the full extent of which cannot be foreseen, and these may be the subject of claim under clauses 12, 13 and 20 of the

111

Conditions. The extent and effect of such events could be mitigated by the employment of specialist supervision under clause 15(3) and by a requirement for the Contractor to include fire prevention and control apparatus, materials and protective clothing under clauses 8(1) and 8(2) of the *Conditions.* The necessity for such precautions would be stated in the 'information supplied', the Schedule and the Specification.

Currently, the following British Standards are applicable to safety precautions in the provision, under clauses 8(2), 19(1) and 19(2), by the Contractor of protective clothing and apparatus for operatives and 'those persons entitled to be upon the Site' (clause 19(1)):

BS 2091: 1969 *Respirators for protection against harmful dust, gases and scheduled agricultural chemicals.*

BS 3776: 1964 *Rescue lines for industrial workers.*

To the authors' knowledge, there have so far been no investigations where radioactive materials have been unexpectedly encountered (that is to say, they have not been publicly reported), but it is certain that this will happen in the future. It would therefore be prudent to include detection apparatus and suitable protective clothing in the equipment required under clauses 8(1) and 8(2) for the investigation of tip sites known to have been formed or added to since 1930 or thereabouts.

British Standards in category 2

Codes which apply to the sampling and testing of soils and materials encountered during ground investigation are also applicable to these *Conditions.* So far, the codes are limited to the sampling, examination and test of soils and rocks, but will be extended to include aspects of, and data relating to manmade materials in additional codes or revision of existing codes, as discussed on p.106.

BS 812, Methods for sampling and testing mineral aggregates, sands and fillers
BS 812 is in four parts:
    Part 1: 1975 *Sampling, size, shape and classification*
    Part 2: 1975 *Physical properties*
    Part 3: 1975 *Mechanical properties*
    Part 4: 1976 *Chemical properties*

This specification should be used in support of the *Conditions* for ground investigation contracts intended to obtain information for the following proposals amongst other applications:

(*a*) Assessment of large areas for mineral working where the quantity, suitability, engineering properties of the aggregate and the ease of working them is to be determined;

(*b*) Investigation of land for borrow-pits within, or within reach of, the construction site where the material may be used for construction purposes.

(*c*) The suitability for construction purposes within the site, of soils to be excavated during construction, such as deep foundations, sub-surface construction, or where large-scale cut and fill work is contemplated.

(*d*) *In situ* drainage, filtration and neutralising properties where disposal of contaminated liquids at great depths below ground level is being considered.

It will be necessary to invoke clause, sub-clause, table, figure or test number when specifying methods of sampling, isolation of representative specimens and the tests applicable. The particular requirements will be included in the Specification and Bill of Quantities. Where there is uncertainty that particular tests will apply, they should be included in the Dayworks Schedule.

British Standard 1377: 1975 Methods of test for soils for civil engineering purposes
This British Standard is the most likely to be used in support of the *Conditions*. Apart from invoking particular tests, it should be used by the Engineer to compare the recommended layout and content of test pro-formas with those used by the soils laboratory (see Chapter 2, p.27 and 30).

It is probable that BS 1377 will be extended to include tests relating to materials found in made ground and contaminated sites, because the engineering properties of these are relevant design data, apart from the potential hazard and the type and degree of contamination caused by, or within, them.

As with the other codes, it will be necessary to invoke specific tests in the Specification.

British Standard 1924: 1975 Methods of test for stablized soils
BS 1924 will be particularly appropriate to ground investigations for road construction projects, including sources of aggregates discussed on p.113. The sampling and testing described is specialised, as are the plant and equipment used in construction. It is therefore necessary for the Engineer, when making his inspection of the laboratories as required by clause 8(3) of the *Conditions* (see Chapter 2), to have this code in mind.

Chapter 9
Settlement of disputes

'Write that down,' said the King to the jury, and
the jury eagerly wrote down all three dates on their
slates, and then added them up, and reduced the
answer to shillings and pence.
Lewis Carroll

Advantages and disadvantages of arbitration

The longest single clause in the *Conditions* is clause 66(1) (the arbitration clause). If either party insists upon a dispute being referred to arbitration, the other party will have difficulty avoiding arbitration, even by the pre-emptive commencement of court proceedings.

Since arbitration is essentially a procedure which demands concensus, co-operation, of the parties to a dispute, in the running of an arbitration is of great importance.

Advantages of arbitration

The principal advantages of arbitration over court proceedings are as follows.

(*a*)   The large majority of disputes under a ground investigation Contract will be technical in origin and nature. By clause 66(1), the parties can agree to the appointment of an arbitrator who has the expertise appropriate to the particular dispute. If the parties cannot agree upon an arbitrator, the President of the Institution of Civil Engineers will appoint as arbitrator a person considered to be appropriate. If the dispute raises geotechnical issues (e.g. clause 12 claims), it is obviously appropriate that an Engineer with sound geotechnical experience should be appointed and it is to be hoped that the parties will agree to the appointment of, or the President will appoint, such a person. Unlike court proceedings, where the parties have no right to select a particular judge, and where few judges have engineering or technical qualifications or experience, parties to arbitration have the benefit of a tribunal of proven or established expertise.

(b) Arbitrations can be concluded much more quickly and at less cost than equivalent court proceedings. There is generally no good reason why any arbitration cannot be concluded within a year of the arbitrator's appointment, whereas court proceedings of a technical nature will take on average a year and a half to two and a half years to reach judgement. The optional Arbitration Procedure 1983 published by the ICE can enable certain types of arbitration disputes to be resolved within a few months.

(c) Arbitration proceedings are more expensive on a *per diem* basis than court proceedings since, in arbitration, the arbitrator's fees and any hire charge for the arbitration rooms have to be borne by either or both of the parties. However, costs can be saved in arbitration by the adoption of informal procedures (for instance, parties need not be represented by lawyers) and by the more expeditious disposal of the whole case.

(d) The Arbitration Act 1979 (which applies to all new arbitrations commenced after 1 August 1979 – see Statutory Instrument SI 1979 No 750) has ensured that arbitrations in England and Wales are less challengeable in the Courts than before. Prior to this Act, the arbitration system had been open to abuse by recalcitrant parties making constant references to the Court; this had the effect of seriously delaying the production of final and enforceable awards. Arbitration awards cannot now be appealed to the Courts unless either:

(a)   the parties agree;

(b)   the Court itself grants leave (Section 1(3)).

The leave of the Court is granted only exceptionally (see Section 1(4) and *Pioneer Shipping Ltd v BTP Dioxide* ('The Nema') 1982 AC 72. The advantage of arbitration is that there is even more finality than in court actions (for which no leave is required to appeal against most final judgements).

(e) An important benefit of arbitration, especially under a ground investigation Contract, is confidentiality and lack of publicity. If an Investigation which gives rise to disputes between the parties to the Contract concerns a politically or locally sensitive development (for instance, large-scale industrial or residential development, nuclear power station or oil terminal sites), the public and publicised venting of such disputes can harm the prospects of the planned development. The effect of confidentiality (which is reinforced by clause 18 of the *Conditions*) is that no publicity can be lawfully provided by the parties, or the arbitrator, of the arbitration or its result; only the parties, their witnesses and lawyers are entitled to attend hearings.

Disadvantages of arbitration
There are, however, a number of disadvantages which parties should bear in mind before embarking upon an arbitration:

(*a*) In arbitration the procedures for dealing with obstructive behaviour by a party are more cumbersome than in court. A claim cannot be dismissed simply because a claimant inexcusably delays proceedings in the arbitration. If a party fails to comply with arbitrator's orders, the arbitrator or other party can apply to the Court for special powers (more akin to the powers of the High Court) to proceed with the reference (Section 5, Arbitration Act 1979).

(*b*) Although many arbitrators are qualified engineers, many have little practical legal training. Time is wasted because some arbitrators, from a desire to be fair to both sides, allow irrelevant matter to be put.

(*c*) The effect of the Arbitration Act 1979 is that it is now much more difficult to appeal against an incorrect award.

Limitation and arbitration

It is a defence to any claim in arbitration that the claim is statute-bound by limitation. Parliament had laid down that any claim pursuant to any contract can be defeated if the arbitration or action was not commenced within six years (for contracts under hand) or 12 years (for contracts under seal) of the date(s) when the cause of action (i.e. the right to claim) first arose (Sections 5 and 8, Limitation Act 1980).

Commencement of arbitration, which has the effect of stopping the limitation periods running, occurs when the party who wishes to proceed serves a written notice on the other party, requiring the particular disputes to be referred to arbitration (Section 34(3), Limitation Act 1980). The notice should also ask that the other party 'concur in the appointment of an arbitrator' (clause 66(1)). Although not compulsory, the written notice is a good opportunity to suggest particular names of potential (albeit independent) arbitrators.

Negligence and the arbitration Clause

The arbitration clause 66 (17) is very widely drawn to cover a wide ambit of dispute:

'If any dispute of any kind whatsoever shall arise between the Employer and the Contractor in connection with or arising out of the Contract or the carrying out of the Investigation. . . .'

Primarily, the clause relates to contractual disputes. However, it is open to the parties to raise claims in negligence as a separate complaint, provided the negligence alleged is in connection with or arises out of the Contract.

The parties (and indeed the Engineer) owe a duty to exercise reasonable care and skill in connection with the performance of the Contract. This duty will be owed to all persons or parties who could foreseeably suffer loss, injury or damage as a result of the failure to exercise reasonable care.

The findings and conclusions of the Contractor in the Investigation and in his Reports may not be acted upon for many years. Since negligence is not actionable until damage caused by the failure to exercise reasonable care has first occurred, a Contractor may be potentially at risk for many years. For example, if in 1984 a Contractor negligently recommends, in his investigation Report, shallow strip foundations founded in peaty material for domestic dwellings, and if the dwellings are constructed in 1993, and begin to subside and crack in 1996, the Contractor could be proceeded against up to the year 2002. The law concerning limitation in actions for negligence has been stated by the House of Lords in Pirelli General Cable Works Limited v Oscar Faber and Partners 1983 2 WLR 6.

Court proceedings

If both parties agree, the dispute can be tried in the Courts. Even if one party does not agree, certain categories of case can be heard in the Courts. For instance, the Court will allow actions to proceed where there is no real dispute between the parties (e.g. outstanding certified sum) or where the Employer seeks to sue both Contractor and Engineer; the Court has a discretion.

Certain specialist judges (official referees) have been assigned in the High Court to deal principally with building and engineering related cases. There are now four permanent official referees of high calibre in London, and other judges who act as official referees hear actions in other parts of England and Wales. Since their judicial business is concentrated on building and engineering disputes, they have built up an expertise and experience in such cases. Indeed, most of the well-known cases in the past 15 years were first tried by official referees (e.g. Pirelli General Cable Works Limited v Oscar Faber and Partners 1983 2 WLR 6).

Commencement of arbitration

It is a precondition to any arbitration under clause 66 that any dispute or difference which has arisen is first referred to the Engineer for his decision. In making this decision, the Engineer must act entirely impartially, although he can invite and hear representations from both sides. For the Engineer to give his decision under clause 66(1), there must first be a dispute or difference; this arises when there is a clear rejection by one party of the claim or argument of the other party (Monmouth CC v Costelloe and Kemple 1965 63 LGR 429, 5 BLR 83).

It is crucial for any party wishing to commence an arbitration that the procedure in clause 66(1) is followed. The Engineer is given three calendar months in which to give his decision. That decision, no matter how wrong it is, will be final and binding on the parties unless either party commences an arbitration within three months of the decision. An anomalous and difficult position arises if the Engineer does not give his decision within three months; the party wishing to proceed to arbitration must then commence the arbitration within six months of the original reference to the Engineer; if he fails to

119

do that, his inaction may effectively have excluded him from having the dispute finally decided by any tribunal.

Arbitration is 'commenced', as explained on p.116, upon the service of written notice requiring the dispute to be referred to arbitration. If given within the requisite three month period after the giving of the Engineer's decision, it will avoid the finality of the Engineer's clause 66 decision, as well as stopping limitation from continuing to run.

Unless the parties agree otherwise, and save in exceptional circumstances (clause 66(2)), arbitration (albeit commenced) cannot proceed before completion of the Investigation. Since the term 'Investigation' is defined to include the Site Operations, Laboratory Testing and the submision of the Report, this may be fairly late on. The Investigation is complete when the Certificate of Completion either is or (if earlier) should have been issued under clause 48.

The arbitrator, whether by agreement of the parties or by nomination of the President of the Institution of Civil Engineers, is employed effectively by contract to the parties. The parties should agree upon terms of payment with the arbitrator. Once this is done, or even if it is not done, the arbitrator can sue either party, upon default, for payment of his agreed or reasonable fees. Conversely, if the arbitrator, having accepted his appointment, fails to issue his award with reasonable dispatch, or is guilty of misconduct, the parties may well be free not to pay for all or part of his services.

Procedure in arbitration

Since arbitration is consensual by nature, the parties may agree upon whatever procedure they like. Accordingly, by agreement, the procedure could be a full-blown court-type hearing or as simple as a mutual exchange of expert geotechnical experts' reports. In the absence of such special agreement, the procedure to be laid down must be such as to bring about the fair and expeditious disposal of the case.

The arbitrator has one principal overriding duty, which is to act fairly as between the parties. This duty is manifested in a number of ways, which include:

(a) Before accepting finally any appointment, the arbitrator should disclose any interest; for instance, if he has acted for or against either party in the past, he should tell the parties, to give them the choice as to whether or not to appoint him.

(*b*) He should at all times give both parties equal and full opportunity to be heard on all questions both of fact and of argument.

(*c*) Save in exceptional circumstances, the arbitrator should hear each party in the presence of the other. He should not entertain argument or evidence from one side which has not been submitted to the other side.

(*d*) Whilst (particularly in arbitrations concerning ground investigation) the arbitrator is entitled to rely upon his own expertise and experience in reaching a decision, he should decide the case upon the evidence placed before him. For instance, a Contractor pursuing a clause 12 claim may present evidence that his boring equipment encountered a mediaeval stone structure; the employer simply denies this; the arbitrator, from his own knowledge, believes that the obstruction could not have been that alleged but was probably unforeseeable rock deposits; theoretically, he should reject the claim; in practice, however, he is entitled informally to tell the parties his views and to invite them to consider their respective positions.

The duty to act fairly does not enable the arbitrator to decide the case on a basis which is contrary to the law. Whilst a judgement of Solomon may be justified on occasion, it is not justified *per se*.

The normal course of events in arbitration is as follows:

(*a*) Confirmation and clarification of appointment.

(*b*) A preliminary meeting is called at which the arbitrator and parties will agree upon the procedure or, failing agreement, the arbitrator may give directions as to how best to have the dispute fairly and expeditiously resolved. Generally, the procedure will require each party in turn to submit in writing their case and defence respectively (pleadings), to make available relevant documents (discovery) and to exchange experts' reports.

(*c*) Up to the time of the hearing, either party or the arbitrator can seek to ensure that the procedure is being adhered to or to impose further directions.

(*d*) At the hearing itself, a court-type procedure is often followed, with the claimants first putting forward their case and witnesses, the respondents being given their chance to cross-examine; the respondents then proceed in the same way, the oral proceedings culminating in final submissions from the parties.

Alternative procedures can be adopted:

(*a*) In certain types of arbitrations, the proceedings can be conducted in writing. For instance, the sole issue might be technical and capable of being resolved by submission to the arbitrator of geotechnical experts' reports. If the arguments are purely legal, written submissions will often suffice.

(*b*) Provided that the parties agree, the Arbitration Procedure issued by the Institution of Civil Engineers (approved February 1983) can be used. This gives the arbitrator much wider powers than would otherwise be available. For instance, the arbitrator may insist that an assessor (legal or technical) sits with him to assist him to assimilate the arguments or evidence (Rule 9).

This document also lays down optional procedures for certain types of dispute which will ensure expeditious and economical disposal of the issues. These are set out in the extracts below:

Part F. Short Procedure
Rule 20. Short Procedure
20.1 Where the parties so agree (either of their own motion or at the invitation of the Arbitrator) the arbitration shall be conducted in accordance with the following *Short Procedure*.
20.2 Each party shall set out his case in the form of a file containing
    (a) a statement as to the orders or awards he seeks
    (b) a statement of his reasons for being entitled to such orders or awards
and (c) copies of any documents on which he relies (including statements) identifying the origin and date of each document
and shall deliver copies of the said file to the other party and to the Arbitrator in such manner and within such time as the Arbitrator may direct.

122

20.3 After reading the parties' cases the Arbitrator may view the site or the Works and may require either or both parties to submit further documents or information in writing.

20.4 Within one calendar month of completing the foregoing steps the Arbitrator shall fix a day when he shall meet the parties for the purpose of

    (a)   receiving any oral submissions which either party may wish to make

and/or (b)   the Arbitrator's putting questions to the parties their representatives or witnesses

For this purpose the Arbitrator shall give notice of any particular person he wishes to question but no person shall be bound to appear before him.

20.5 Within one calendar month following the conclusion of the meeting under Rule 20.4 or such further period as the Arbitrator may reasonably require the Arbitrator shall make and publish his Award.

Part G. Special Procedure for Experts

Rule 22. Special Procedure for Experts

22.1 Where the parties so agree (either of their own motion or at the invitation of the Arbitrator) the hearing and determination of any issues of fact which depend upon the evidence of experts shall be conducted in accordance with the following *Special Procedure*.

22.2 Each party shall set out his case on such issues in the form of a file containing

    (a)   a statement of the factual findings he seeks

    (b)   a report or statement from and signed by each expert upon whom that party relies

and (c)   copies of any other documents referred to in each expert's report or statement or on which the party relies identifying the origin and date of each document

and shall deliver copies of the said file to the other party and to the Arbitrator in such manner and within such time as the Arbitrator may direct.

22.3 After reading the parties' cases the Arbitrator may view the site or the Works and may require either or both parties to submit further documents or information in writing.

22.4 Thereafter the Arbitrator shall fix a day when he shall meet the experts whose reports or statements have been submitted. At the meeting each expert may address the Arbitrator and put questions to any other expert representing the other party. The Arbitrator shall so direct the meeting as to ensure that each expert has an adequate opportunity to explain his opinion and to comment upon any opposing opinion. No other person shall be entitled to address the Arbitrator or question any expert unless the parties and the Arbitrator so agree.

22.5 Thereafter the Arbitrator may make and publish an award setting out with such details or particulars as may be necessary his decision upon the issues dealt with.

Expert evidence

In any legal proceedings, evidence is normally restricted to first-hand observation of events. Expert evidence falls into a different category in that it is opinion or comment on established facts.

In ground investigation arbitrations, the most common type of case requiring expert witnesses will be clause 12 (unforeseeable conditions or obstructions) and clause 13 (physical impossibility) claims. Parties must exercise care in the selection of expert witnesses to ensure that the discipline, experience and expertise of the expert selected is appropriate. The importance of this is such that the following is provided for the guidance of those engineers so engaged.

The role of the expert witness

When first chosen to investigate a claim and give his opinion on matters in dispute, an engineer may feel some flush of pride at being referred to as an 'expert witness'. However, this is merely the legal term for a technical witness and does not imply that the person concerned possesses the ultimate in geotechnical expertise.

The function of the expert witness in relation to forensic investigation planned and undertaken personally is much more complicated than the presentation of an opinion on facts and data determined by others. There are, however, basic requirements common to the presentation of expert evidence and those of major importance are listed here.

(*a*) By accepting an appointment, the potential witness must determine whether the matter is within his sphere of competence and, partly to this end, will require as much detail as is available regarding the contract documents and all the events leading to the dispute.

(*b*) He should visit the Site in company with the party whom he is to represent and, if further data is necessary to support the case, will require it to be obtained by that client or by his own resources.

(*c*) He will make it clear that his evidence and opinions will be independent and not be biased in favour of the client, and, although he will seek to establish favourable factors, his evidence will be truth based on his own interpretation of the facts.

(*d*) Where possible, his evidence should be in sequence of occurrence, supported by photographs, illustrations and, if applicable, exhibits of materials involved.

Given this, the overriding factor is the independent expert witness's understanding that his job is primarily to assist the judge or arbitrator to reach a just award.

124

Finality of award

Clause 66(1) expressly declares that the award of the arbitrator shall be final and binding upon the parties. Save in exceptional circumstances, the award is not challengeable. Any award by an arbitrator in England and Wales not only is inherently enforceable, but can also be enforced in the Courts.

Challenging awards

There are four main methods of challenging Arbitrator's awards:

(a) Where the parties have expressly agreed there should be a right to appeal to the Courts.

(b) Where the parties have not expressly agreed, a party may appeal to the (High) Court on any question of law arising on the award provided that the Court gives leave. Leave will only be given when the question of law would substantially affect the rights of the parties (Section 1(3) of the Arbitration Act 1979). Leave is granted exceptionally.

(c) Where the arbitrator 'misconducts' himself; in cases of misconduct, the award can be set aside by the Court. 'Misconduct' is normally any serious failure on the part of the arbitrator to act fairly towards the parties. Examples in ground investigation arbitrations could be where an arbitrator allows one party to submit a geotechnical report without giving the other party the opportunity to be heard; or where a conscientious arbitrator doing his 'homework' on his own discovers that the geological maps contain vital information not referred to by either side and then bases his award on that information without giving the parties a chance to be heard on the point.

(d) Where the arbitrator exceeds his jurisdiction; for instance, if the arbitrator decides a dispute which has not been referred to him, the award can effectively be challenged.

125

Appendix 1
ICE Conditions of Contract
for Ground Investigation

CONDITIONS OF CONTRACT

DEFINITIONS AND INTERPRETATION

Definitions **1.** (1) In the Contract (as hereinafter defined) the following words and expressions shall have the meanings hereby assigned to them except where the context otherwise requires:

(a) 'Employer' means . . . .
of . . . .
and includes the Employer's personal representatives or successors;

(b) 'Contractor' means the person or persons firm or company whose tender has been accepted by the Employer and includes the Contractor's personal representatives successors and permitted assigns;

(c) 'Engineer' means . . . .
or other the Engineer appointed from time to time by the Employer and notified in writing to the Contractor to act as Engineer for the purposes of the Contract in place of said . . . .

(d) 'Engineer's Representative' means a person being the resident engineer or assistant of the Engineer or clerk of works appointed from time to time by the Employer or the Engineer and notified in writing to the Contractor by the Engineer to perform the functions set forth in Clause 2(1);

(e) 'Contract' means the Conditions of Contract Specification Schedules Drawings Priced Bill of Quantities the Tender the written acceptance thereof and the Contract Agreement (if completed);

(f) 'Specification' means the specification referred to in the Tender and any modification thereof or addition thereto as may from time to time be furnished or approved in writing by the Engineer;

(g) 'Drawings' means the drawings referred to in the Specification and any modification of such drawings approved in writing by the Engineer and such other drawings as may from time to time be furnished or approved in writing by the Engineer;

(h) 'Schedules' means the schedules and lists of Site Operations Laboratory Testing and other requirements referred to in the Specification;

(i) 'Tender Total' means the total of the Priced Bill of Quantities at the date of acceptance of the Contractor's Tender for the Investigation;

(j) 'Contract Price' means the sum to be ascertained and paid in accordance with the provisions hereinafter contained for carrying out the Investigation in accordance with the Contract;

(k) 'Site Operations' means all the work of every kind including Ancillary Works required to be carried out on under in or through the Site in accordance with the Contract;

(l) 'Ancillary Works' means all appliances or things of whatsoever nature required to be installed or constructed on under in or through the Site and which are to remain on Site and become the property of the Employer in accordance with the Contract upon the issue of a Certificate of Completion in respect of the Site operations or section or part thereof as the case may be;

126

(*m*) 'Laboratory Testing' means the testing operations and processes necessary for the preparation of the Report to be carried out in accordance with the Contract at a laboratory approved by the Engineer on samples and cores obtained during the Site Operations;

(*n*) 'Report' means the report to be prepared and submitted in accordance with the Contract;

(*o*) 'Investigation' means the Site Operations together with the Laboratory Testing and Report preparation and submission;

(*p*) 'Section' means a part of the Investigation separately identified in the Appendix to the Form of Tender;

(*q*) 'Site' means the lands and other places on under in or through which the Site Operations are to be executed and any lands places or access thereto provided by the Employer for the purposes of the Contract;

(*r*) 'Equipment' means any appliances or things of whatsoever nature required temporarily for carrying out the Site Operations but does not include anything which forms part of the Ancillary Works.

(2)   Words importing the singular also include the plural and vice versa where the context requires.   **Singular and Plural**

(3)   The headings and marginal notes in the Conditions of Contract shall not be deemed to be part thereof or be taken into consideration in the interpretation or construction thereof or of the Contract.   **Headings and Marginal Notes**

(4)   All references herein to clauses are references to clauses numbered in the Conditions of Contract and not to those in any other document forming part of the Contract.   **Clause References**

(5)   The word 'cost' when used in the Conditions of Contract shall be deemed to include overhead costs whether on or off the Site except where the contrary is expressly stated.   **Cost**

ENGINEER'S REPRESENTATIVE

2.   (1)   The functions of the Engineer's Representative are to watch and supervise the Investigation. He shall have no authority to relieve the Contractor of any of his duties or obligations under the Contract nor except as expressly provided hereunder to order any work involving delay or any extra payment by the Employer nor to make any variation of or in the Investigation.   **Functions and Powers of Engineer's Representative**

(2)   The Engineer or the Engineer's Representative may appoint any number of persons to assist the Engineer's Representative in the exercise of his functions under sub-clause (1) of this Clause. He shall notify to the Contractor the names and functions of such persons. The said assistants shall have no power to issue any instructions to the Contractor save insofar as such instructions may be necessary to enable them to discharge their functions and to secure their acceptance of methods materials workmanship or Ancillary Works as being in accordance with the Specification Drawings and Schedules and any instructions given by any of them for those purposes shall be deemed to have been given by the Engineer's Representative.   **Appointment of Assistants**

(3)   The Engineer may from time to time in writing authorise the Engineer's Representative or any other person responsible to the Engineer to act on behalf of the Engineer either generally in respect of the Contract or specifically in respect of particular Clauses of these Conditions of Contract and any act of any such person within the scope of this authority shall for the purposes of the contract constitute an act of the Engineer. Prior notice in writing of any such authorisation shall be given by the Engineer to the Contractor. Such authorisation shall continue in force until such time as the Engineer shall notify the Contractor in writing that the same is determined. Provided that such authorisation shall not be given in respect of any decision to be taken or certificate to be issued under Clauses 12(3) 44 48 60(3) 61 63 and 66.   **Delegation by Engineer**

(4)   If the Contractor shall be dissatisfied by reason of any instruction of any assistant of the Engineer's Representative duly appointed under sub-clause (2) of this Clause he shall be entitled to refer the matter to the Engineer's Representative who shall thereupon confirm reverse or vary such instruction. Similarly if the Contractor shall be dissatisfied by reason of any act of the Engineer's Representative or other person duly authorised by the Engineer under sub-clause (3) of this Clause he shall be entitled to refer the matter to the Engineer for his decision.   **Reference to Engineer or Engineer's Representative**

ASSIGNMENT AND SUB-LETTING

3.   The Contractor shall not assign the Contract or any part thereof or any benefit or interest therein or thereunder without the written consent of the Employer.   **Assignment**

4.   The Contractor shall not sub-let the whole of the Investigation. Except where otherwise provided by the Contract the Contractor shall not sub-let any part of the Investigation without the written consent of the Engineer and such consent if given shall not relieve the Contractor from any liability or obligation under the Contract and he shall be responsible for the acts defaults and neglects of any sub-contractor his agents servants or workmen as fully as if they were the acts   **Sub-letting**

defaults or neglects of the Contractor his agents servants or workmen. Provided always that the provision of labour on a piece-work basis shall not be deemed to be a sub-letting under this Clause.

CONTRACT DOCUMENTS

Documents Mutually Explanatory

5. The several documents forming the Contract are to be taken as mutually explanatory of one another and in case of ambiguities or discrepancies the same shall be explained and adjusted by the Engineer who shall thereupon issue to the Contractor appropriate instructions in writing which shall be regarded as instructions issued in accordance with Clause 13.

Supply of Documents

6. Upon acceptance of the Tender two copies of the Drawings and Schedules referred to in the Specification and of the Conditions of Contract the Specification and (unpriced) Bill of Quantities shall be furnished to the Contractor free of charge. Copyright of the Drawings Schedules and Specification and of the Bill of Quantities (except the pricing thereof) shall remain in the Engineer but the Contractor may obtain or make at his own expense any further copies required by him. At the completion of the Contract the Contractor shall return to the Engineer all Drawings Schedules and the Specification whether provided by the Engineer or obtained or made by the Contractor.

Reference Points

7. (1) The Engineer shall designate or establish bench marks and on-site reference points to enable the Contractor to set out the Site Operations in accordance with Clause 17.

Further Drawings and Instructions

(2) The Engineer shall have full power and authority to supply and shall supply to the Contractor from time to time during the progress of the Investigation such modified or further drawings schedules or instructions as shall in the Engineer's opinion be necessary for the purpose of the proper and adequate execution of the Investigation and the Contractor shall carry out and be bound by the same.

Notice by Contractor

(3) The Contractor shall give adequate notice in writing to the Engineer of any further drawing schedule or specification that the Contractor may require for the execution of the Investigation or otherwise under the Contract.

Delay in Issue

(4) If by reason of any failure or inability of the Engineer to issue at a time reasonable in all the circumstances drawings schedules or instructions requested by the Contractor and considered necessary by the Engineer in accordance with sub-clause (1) of this Clause or failure by the Engineer to establish bench-marks or on-site reference points the Contractor suffers delay or incurs cost then the Engineer shall take such delay into account in determining any extension of time to which the Contractor is entitled under Clause 44 and the Contractor shall subject to Clause 52(4) be paid in accordance with Clause 60 the amount of such cost as may be reasonable. If such drawings schedules or instructions require any variation to any part of the Investigation the same shall be deemed to have been issued pursuant to Clause 51.

One Copy of Documents to be kept on site

(5) One copy of the Drawings Schedules Specification and instructions furnished to the Contractor as aforesaid shall be kept by the Contractor on the Site and the same shall at all reasonable times be available for inspection and use by the Engineer and the Engineer's Representative and by any other person authorised by the Engineer in writing.

GENERAL OBLIGATIONS

Contractor's General Responsibilities

8. (1) The Contractor shall subject to the provisions of the Contract carry out the Investigation and provide all labour materials Equipment instrumentation transport to and from and in or about the Site and everything whether of a temporary or permanent nature required in and for the Investigation so far as the necessity for providing the same is specified in or reasonably to be inferred from the Contract.

Safety of Site Operations

(2) The Contractor shall take full responsibility for the adequacy stability and safety of all Site Operations Laboratory Testing and methods of working.

Suitability of Laboratories

(3) The Contractor shall submit to the Engineer the name and address of the laboratories undertaking Laboratory Testing in accordance with the Contract and shall obtain the Engineer's approval in writing before the services and equipment of such laboratories may be used in the execution of the Contract.

Design Responsibility

(4) Except as may be expressly provided in the Contract the Contractor shall not be responsible for the design or specification of any Ancillary Works required to be installed or constructed in accordance with the Contract.

Contract Agreement

9. The Contractor shall when called upon so to do enter into and execute a Contract Agreement (to be prepared at the cost of the Employer) in the form annexed.

Sureties

10. If the Tender shall contain an undertaking by the Contractor to provide when required two good and sufficient sureties or to obtain the guarantee of an Insurance Company or Bank to be jointly and severally bound with the Contractor in a sum not exceeding 10 per cent of the Tender Total for the due performance of the Contract under the terms of a Bond the said sureties Insurance Company or Bank and the terms of the said Bond shall be such as shall be approved by the Employer and the provision of such sureties or the obtaining of such guarantee and the cost of the Bond to be so entered into shall be at the expense in all respects of the Contractor unless the Contract otherwise provides. Provided always that if the form of Bond approved by the Employer shall contain provisions for the determination by an arbitrator of any dispute or difference concerning the relevant date for the discharge of the Sureties'/Surety's obligations under the said Bond:

(a) the Employer shall be deemed to be a party to the said Bond for the purpose of doing all things necessary to carry such provisions into effect

(b) any agreement decision award or other determination touching or concerning the relevant date for the discharge of the Sureties'/Surety's obligations under the said Bond shall be wholly without prejudice to the resolution or determination of any dispute or difference between the Employer and the Contractor pursuant to the provisions of Clause 66.

11. (1) The Contractor shall be deemed to have inspected and examined the Site and its surroundings and to have informed himself before submitting his tender as to the general nature of the geology (so far as is practicable and having taken into account any information in connection therewith which may have been provided by or on behalf of the Employer) the form and nature of the Site the extent and nature of the work and materials necessary for the completion of the Investigation the means of communication with and access to the Site the accommodation he may require and in general to have obtained for himself all necessary information (subject as above-mentioned) as to risks contingencies and all other circumstances influencing or affecting his tender. **Inspection of Site**

(2) The Contractor shall be deemed to have satisfied himself before submitting his tender as to the correctness and sufficiency of the rates and prices stated by him in the Priced Bill of Quantities which shall (except insofar as it is otherwise provided in the Contract) cover all his obligations under the Contract. **Sufficiency of Tender**

12. (1) If during the execution of the Site Operations the Contractor shall encounter physical conditions (other than weather conditions or conditions due to weather conditions) or artificial obstructions which conditions or obstructions he considers could not reasonably have been foreseen by an experienced contractor and the Contractor is of opinion that additional cost will be incurred which would not have been incurred if the physical conditions or artificial obstructions had not been encountered he shall if he intends to make any claim for additional payment give notice to the Engineer pursuant to Clause 52(4) and shall specify in such notice the physical conditions and/or artificial obstructions encountered and with the notice if practicable or as soon as possible thereafter give details of the anticipated effects thereof the measures he is taking or is proposing to take and the extent of the anticipated delay in or interference with the execution of the Investigation. **Adverse Physical Conditions and Artificial Obstructions**

(2) Following receipt of a notice under sub-clause (1) of this Clause the Engineer may if he thinks fit *inter alia*: **Measures to be Taken**

(a) require the Contractor to provide an estimate of the cost of the measures he is taking or is proposing to take;

(b) approve in writing such measures with or without modification;

(c) give written instructions as to how the physical conditions or artificial obstructions are to be dealt with;

(d) order a suspension under Clause 40 or a variation under Clause 51.

(3) To the extent that the Engineer shall decide that the whole or some part of the said physical conditions or artificial obstructions could not reasonably have been foreseen by an experienced contractor the Engineer shall take any delay suffered by the Contractor as a result of such conditions or obstructions into account in determining any extension of time to which the Contractor is entitled under Clause 44 and the Contractor shall subject to Clause 52(4) (notwithstanding that the Engineer may not have given any instructions or orders pursuant to sub-clause (2) of this Clause) be paid in accordance with Clause 60 such sum as represents the reasonable cost of carrying out any additional work done and additional Equipment used which would not have been done or used had such conditions or obstructions or such part thereof as the case may be not been encountered together with a reasonable percentage addition thereto in respect of profit and the reasonable costs incurred by the Contractor by reason of any unavoidable delay or disruption of working suffered as a consequence of encountering the said conditions or obstructions or such part thereof. **Delay and Extra Cost**

(4) If the Engineer shall decide that the physical conditions or artificial obstructions could in whole or in part have been reasonably foreseen by an experienced contractor he shall so inform the Contractor in writing as soon as he shall have reached that decision but the value of any variation previously ordered by him pursuant to sub-clause (2)(d) of this Clause shall be ascertained in accordance with Clause 52 and included in the Contract Price. **Conditions Reasonably Foreseeable**

13. (1) Save insofar as it is legally or physically impossible the Contractor shall carry out the Investigation in strict accordance with the Contract to the satisfaction of the Engineer and shall comply with and adhere strictly to the Engineer's instructions and directions on any matter connected **Work to be to Satisfaction of Engineer**

therewith (whether mentioned in the Contract or not). The Contractor shall take instructions and directions only from the Engineer or (subject to the limitations referred to in Clause 2) from the Engineer's Representative.

Mode and Manner of Execution

(2)　The whole of the labour materials Equipment and instrumentation to be provided by the Contractor under Clause 8 and the mode manner and speed of carrying out the Investigation are to be of a kind and conducted in a manner approved of by the Engineer.

Delay and Extra Cost

(3)　If in pursuance of Clause 5 or sub-clause (1) of this Clause the Engineer shall issue instructions or directions which involve the Contractor in delay or disrupt his arrangements or methods of working so as to cause him to incur cost beyond that reasonably to have been foreseen by an experienced contractor at the time of tender then the Engineer shall take such delay into account in determining any extension of time to which the Contractor is entitled under Clause 44 and the Contractor shall subject to Clause 52(4) be paid in accordance with Clause 60 the amount of such cost as may be reasonable. If such instructions or directions require any variation to any part of the Investigation the same shall be deemed to have been given pursuant to Clause 51.

Continuation Required by Conditions

(4)　If during the execution of Site Operations the Contractor shall encounter ground or geological conditions which in his opinion make it necessary for the effectiveness of the Investigation or for the adequacy of the Report to continue the operations of boring drilling excavation sampling or in situ testing to a greater depth than is included in the Schedules before Equipment is moved from the position of the borehole drill hole or excavation and the Engineer's or (subject to the limitations referred to in Clause 2) the Engineer's Representative's instructions cannot be immediately obtained the Contractor may continue such operations or change the mode of operation at his own discretion provided always that the cost of such continued operations or changed mode of operation shall not exceed such sum as may have been agreed between the Engineer and the Contractor in writing at the commencement of the Site Operations unless a further instruction or specification shall have been subsequently issued by the Engineer in accordance with sub-clause (5) of the Clause.

Engineer's Instructions

(5)　In the event of operations being continued in accordance with sub-clause (4) of this Clause every endeavour shall simultaneously be made to obtain the Engineer's or (subject to the limitations referred to in Clause 2) the Engineer's Representative's instructions.

Justified Continuation

(6)　Subject to any instruction of the Engineer previously given pursuant to this sub-clause such continued or changed operations shall be deemed to have been carried out as a variation ordered pursuant to Clause 51, provided always that if the Engineer shall decide that the Contractor was not justified in continuing or changing the operation in accordance with sub-clause (4) of this Clause the Engineer shall so inform the Contractor in writing and shall issue further instructions or specification as he may think fit to govern how any similar ground or geological conditions which may be encountered in the future shall be dealt with.

Backfilling of Boreholes

(7)　On completion of a borehole to its specified depth or to a greater depth under sub-clause (4) of this Clause the Contractor may unless the contract provides otherwise backfill the borehole without further reference to the Engineer or the Engineer's Representative.

Programme to be Furnished

14.　(1)　Within 21 days after the acceptance of his Tender the Contractor shall submit to the Engineer for his approval a programme showing the order in which he proposes to carry out the Investigation and thereafter shall furnish such further details and information as the Engineer may reasonably require in regard thereto. The Contractor shall at the same time also provide in writing for the information of the Engineer a description of the arrangements and methods which the Contractor proposes to adopt for the carrying out of the Investigation. The programme submitted by the Contractor pursuant to this sub-clause shall take into account the period or periods for completion of the Investigation or different Sections thereof and the periods required by the Engineer for approval of testing schedules and reports which are provided for in the Appendix to the Form of Tender.

Revision of Programme

(2)　Should it appear to the Engineer at any time that the actual progress of the Investigation does not conform with the approved programme referred to in sub-clause (1) of this Clause the Engineer shall be entitled to require the Contractor to produce a revised programme showing the modifications to the original programme necessary to ensure completion of the Investigation or any Section within the time for completion as defined in Clause 43 or extended time granted pursuant to Clause 44(2).

Engineer's Consent

(3)　The Engineer shall inform the Contractor in writing within a reasonable period after receipt of the information submitted in accordance with sub-clause (1) of this Clause either:

(*a*)　that the Contractor's proposed methods have the consent of the Engineer; or
(*b*)　in what respects in the opinion of the Engineer they fail to meet the requirements of the Drawings or Specification.

In the latter event the Contractor shall take such steps or make such changes in the said methods as may be necessary to meet the Engineer's requirements and to obtain his consent. The Contractor shall not change the methods which have received the Engineer's consent without the further consent in writing of the Engineer which shall not be unreasonably withheld.

(4) If the Engineer's consent to the proposed methods of working shall be unreasonably delayed or if the requirements of the Engineer pursuant to sub-clause (3) of this Clause or if a change in the sequence of operations required by the Engineer could not reasonably have been foreseen by an experienced contractor at the time of tender and if in consequence of any of the aforesaid the Contractor unavoidably incurs delay or cost the Engineer shall take such delay into account in determining any extension of time to which the Contractor is entitled under Clause 44 and the Contractor shall subject to Clause 52(4) be paid in accordance with Clause 60 such sum in respect of the cost incurred as the Engineer considers fair in all the circumstances.

Delay and Extra Cost

(5) If the time taken by the Engineer in giving his approval to the Contractor's proposed testing schedule or the draft Report or the final Report shall exceed the appropriate Period for Approval shown in the Appendix to the Form of Tender and as a result the time for completion of the Investigation or any Section thereof shall be exceeded the Engineer shall grant an extension of time in accordance with Clause 44 in respect of the time taken by him in excess of the appropriate Period for Approval shown in the Appendix to the Form of Tender which Period for Approval shall be calculated from the time elapsed between despatch of the Contractor's proposed testing schedule or the draft Report or the final Report as appropriate and receipt by the Contractor of the Engineer's respective disapproval or approval.

Period for Approvals

(6) Approval by the Engineer of the Contractor's programme in accordance with sub-clauses (1) and (2) of this Clause and the consent of the Engineer to the Contractor's proposed methods of working in accordance with sub-clause (3) of this Clause shall not relieve the Contractor of any of his duties or responsibilities under the Contract.

Responsibility Unaffected by Approval

15. (1) The Contractor shall give or provide all necessary superintendence during the execution of the Investigation. Such superintendence shall be given by sufficient persons being suitably qualified having adequate experience and knowledge of the operations to be carried out (including the methods and techniques required the hazards likely to be encountered and the prevention of accidents) as may be requisite for the satisfactory execution of the Investigation. The Contractor shall be responsible for the safety of all operations.

Contractor's Superintendence

(2) The Contractor shall provide a competent suitably qualified and authorised agent or representative approved of in writing by the Engineer (which approval may at any time be withdrawn). Such authorised agent or representative shall be in full charge of the Investigation and shall receive on behalf of the Contractor directions and instructions from the Engineer or (subject to the limitations of Clause 2) the Engineer's Representative. Such agent or representative may appoint a person who shall be constantly on the Site during the Site Operations and shall receive instructions relating to the Site Operations.

Contractor's Agent

(3) If in addition to the superintendence in accordance with sub-clauses (1) and (2) of this Clause the Contract shall require or the Engineer direct the Contractor to make available on the Site or elsewhere the services of suitably qualified persons for description of soils and rocks logging of trial pits execution of geological and geotechnical appraisals other technical and advisory services and the preparation of technical reports the extent and scope of the service required shall be specified in the Contract.

Services of a Specialist

16. The Contractor shall employ or cause to be employed in and about the execution of the Investigation and in the superintendence thereof only such persons as are careful skilled and experienced in their several trades and callings and the Engineer shall be at liberty to object to and require the Contractor to remove from the Investigation any person employed by the Contractor in or about the Investigation who in the opinion of the Engineer misconducts himself or is incompetent or negligent in the performance of his duties or fails to conform with any particular provisions with regard to safety which may be set out in the Specification or persists in any conduct which is prejudicial to safety or health and such persons shall not be again employed upon the Investigation without the permission of the Engineer.

Removal of Contractor's Employees

17. The Contractor shall use the bench marks and on-site reference points established by the Engineer pursuant to Clause 7(1) to determine and record levels and set out the positions of the Site Operations. The Contractor shall be responsible for the true and proper setting-out of the Site Operations and for the correctness of the position levels dimensions and alignment of all parts of the Site Operations and for the provision of all necessary instruments appliances and labour in connection therewith. If at any time during the progress of the Investigation any error shall appear or arise in the position levels dimensions or alignment of any part of the Site Operations the Contractor on being required so to do by the Engineer shall at his own cost rectify such error to the satisfaction of the Engineer unless such error is based on incorrect or inaccurate bench-marks on-site reference points or incorrect data supplied in writing by the Engineer or the Engineer's Representative in which case the cost of rectifying the same shall be borne by the Employer. The checking of any setting-out or of any line or level by the Engineer or the Engineer's Representative shall not in

Setting-out

131

any way relieve the Contractor of his responsibility for the correctness thereof and the Contractor shall carefully protect and preserve all bench-marks sight rails pegs and other things used in setting out the Site Operations.

Confidentiality and Security

18. The Contractor shall at all times keep confidential between himself the Engineer and the Employer all information obtained from the carrying out of the Contract and no information relating thereto shall be communicated in any form to any person or body other than those named in written authorisation given by the Employer or the Engineer. The Contractor shall as far as is reasonably practicable prevent the entry of unauthorised persons to the Site to examine the work in progress and safeguard and secure samples taken from the borings or trial pits and records against unauthorised examination.

Safety and Security

19. (1) The Contractor shall throughout the progress of the Site Operations have full regard for the safety of all persons entitled to be upon the Site and shall keep the Site (so far as the same is under his control) and the Site Operations (except insofar as the same consist of Ancillary Works covered by a Certificate of Completion or which have been delivered up to the Employer) in an orderly state appropriate to the avoidance of danger to such persons and shall *inter alia* in connection with the Site Operations provide and maintain at his own cost all lights guards fencing warning signs and watching when and where necessary or required by the Engineer or by any competent statutory or other authority for the protection of the Site Operations or for the safety and convenience of the public or others.

Employer's Responsibilities

(2) If under Clause 31 the Employer shall carry out work on the Site with his own workmen he shall in respect of such work:

(*a*) have full regard to the safety of all persons entitled to be upon the Site; and

(*b*) keep the Site in an orderly state appropriate to the avoidance of danger to such persons.

If under Clause 31 the Employer shall employ other contractors on the Site he shall require them to have the same regard for safety and avoidance of danger.

Care of the Investigation

20. (1) The Contractor shall take full responsibility for the care of the Site Operations from the date of the commencement thereof until 14 days after the Engineer shall have issued a Certificate of Completion for the whole of the Site Operations pursuant to Clause 48. Provided that if the Engineer shall issue a Certificate of Completion in respect of any Section or part of the Site Operations before he shall issue a Certificate of Completion in respect of the whole of the Site Operations the Contractor shall cease to be responsible for the care of the Site Operations and the Ancillary Works comprised within that Section or part 14 days after the Engineer shall have issued the Certificate of Completion in respect of that Section or part and the responsibility for the care of the Site relating to that Section or part and of any Ancillary Works included in that Section or part shall thereupon pass to the Employer. Provided further that the Contractor shall take full responsibility for the care of any outstanding work which he shall have undertaken to finish during the Period of Maintenance until such outstanding work is complete.

Care of Samples and Cores

(2) The Contractor shall unless it is otherwise provided for in the Contract take full responsibility for the care and storage of the samples and cores obtained from the investigation at his cost until 28 days after the issue of the Report to the Engineer or in the case of a phased investigation the relevant section of the Report. After the said period the Contractor shall give the Engineer 14 days written notice of his intention to charge rental for the storage of the cores and samples. The Engineer shall then either give instructions for the immediate disposal for the samples and cores at the Contractor's expense or state his storage or other requirements giving an indication of his programme. The rental charged for the cores and samples stored shall commence after the expiry of the said 14 days notice and shall continue until the Engineer gives disposal instructions.

Responsibility for Reinstatement

(3) In case any damage loss or injury from any cause whatsoever (save and except the Excepted Risks as defined in sub-clause (4) of this Clause) shall happen to the Site Operations the samples and cores or the Report or any part thereof while the Contractor shall be responsible for the care thereof the Contractor shall at his own cost repair make good or replace the same so that at completion the Ancillary Works shall be in good order and condition and the Investigation shall be in conformity in every respect with the requirements of the Contract and the Engineer's instructions. To the extent that any such damage loss or injury arises from any of the Excepted Risks the Contractor shall if required by the Engineer repair make good or replace the same as aforesaid at the expense of the Employer. The Contractor shall also be liable for any damage to the Ancillary Works occasioned by him in the course of any operations carried out by him for the purpose of completing any outstanding work or for complying with his obligations under Clauses 49 and 50.

Excepted Risks

(4) The 'Excepted Risks' are riot war invasion act of foreign enemies hostilities (whether war be declared or not) civil war rebellion revolution insurrection or military or usurped power ionising radiations or contamination by radioactivity from any nuclear fuel or from any nuclear waste from the combustion of nuclear fuel radioactive toxic explosive or other hazardous properties of any explosive nuclear assembly or nuclear component thereof pressure waves caused by aircraft or other

aerial devices travelling at sonic or supersonic speeds or a cause due to use or occupation by the Employer his agents servants or other contractors (not being employed by the Contractor) of any part of the Site Operations or to fault defect error or omission in the design of the Investigation (other than a design provided by the Contractor pursuant to his obligations under the Contract).

21. Without limiting his obligations and responsibilities under Clause 20 the Contractor shall insure in the joint names of the Employer and the Contractor: **Insurance, etc.**

    (*a*)   the Investigation (including for the purposes of this Clause any unfixed materials or other things delivered to the Site for incorporation therein) to its full value;

    (*b*)   the Equipment to its full value;

    (*c*)   the cores samples and test results records and results obtained and made in the course of the Investigation to an amount equivalent to the cost of their replacement;

against all loss or damage from whatever cause arising (other than the Excepted Risks) for which he is responsible under the terms of the Contract and in such manner that the Employer and Contractor are covered for the period stipulated in Clause 20(1) and are also covered for loss or damage arising during the Period of Maintenance from such cause occurring prior to the commencement of the Period of Maintenance and for any loss or damage occasioned by the Contractor in the course of any operation carried out by him for the purpose of complying with his obligations under Clauses 49 and 50.

Provided that without limiting his obligations and responsibilities as aforesaid nothing in this Clause contained shall render the Contractor liable to insure against the necessity for the repair or reconstruction or repetition of any work constructed with materials and workmanship not in accordance with the requirements of the Contract unless the Bill of Quantities shall provide a special item for this insurance.

Such insurances shall be effected with an insurer and in terms approved by the Employer (which approval shall not be unreasonably withheld) and the Contractor shall whenever required produce to the Employer the policy or policies of insurance and the receipts for payment of the current premiums.

22. (1) The Contractor shall (except if and so far as the Contract otherwise provides) indemnify and keep indemnified the Employer against all losses and claims for injuries or damage to any person or property whatsoever (other than the Investigation for which insurance is required under Clause 21 but including surface or other damage to land being the Site suffered by any persons in beneficial occupation of such land) which may arise out of or in consequence of the Investigation and against all claims demands proceedings damages costs charges and expenses whatsoever in respect thereof or in relation thereto. Provided always that: **Damage to Persons and Property**

    (*a*)   the Contractor's liability to indemnify the Employer as aforesaid shall be reduced proportionately to the extent that the act or neglect of the Employer his servants or agents may have contributed to the said loss injury or damage;

    (*b*)   nothing herein contained shall be deemed to render the Contractor liable for or in respect of or to indemnify the Employer against any compensation or damages for or with respect to:

       (i)   damage to crops being on the Site (save insofar as possession has not been given to the Contractor);

       (ii)   the use or occupation of land (which has been provided by the Employer) for the purpose of carrying out Site Operations (including consequent losses of crops) or interference whether temporary or permanent with any right of way light air or water or other easement or quasi easement which are the unavoidable result of the Site Operations carried out in accordance with the Contract;

       (iii)   the right of the Employer to have the Site Operations or any part thereof carried out on over under in or through any land;

       (iv)   damage which is the unavoidable result of the Site Operations in accordance with the Contract;

       (v)   injuries or damage to persons or property resulting from any act or neglect or breach of statutory duty done or committed by the Engineer or the Employer his agents servants or other contractors (not being employed by the Contractor) or for or in respect of any claims demands proceedings damages costs charges and expenses in respect thereof or in relation thereto.

(2) The Employer will save harmless and indemnify the Contractor from and against all claims demands proceedings damages costs charges and expenses in respect of the matters referred to in the proviso to sub-clause (1) of this Clause. Provided always that the Employer's liability to indemnify the Contractor under paragraph (v) of proviso (b) to sub-clause (1) of this Clause shall be reduced proportionately to the extent that the act or neglect of the Contractor or his subcontractors servants or agents may have contributed to the said injury or damage. **Indemnity by Employer**

Insurance against Damage to Persons and Property

23. (1) Throughout the execution of the Investigation the Contractor (but without limiting his obligations and responsibilities under Clause 22) shall insure against any damage loss or injury which may occur to any property or to any person by or arising out of the execution of the Investigation or in the carrying out of the Contract otherwise than due to the matters referred to in proviso (b) to Clause 22 (1).

Amount and Terms of Insurance

(2) Such insurance shall be effected with an insurer and in terms approved by the Employer (which approval shall not be unreasonably withheld) and for at least the amount stated in the Appendix to the Form of Tender. The terms shall include a provision whereby in the event of any claim in respect of which the Contractor would be entitled to receive indemnity under the policy being brought or made against the Employer the insurer will indemnify the Employer against such claims and any costs charges and expenses in respect thereof. The Contractor shall whenever required produce to the Employer the policy or policies of insurance and the receipts for payment of the current premiums.

Accident or Injury to Workmen

24. The Employer shall not be liable for or in respect of any damages or compensation payable at law in respect or in consequence of any accident or injury to any workman or other person in the employment of the Contractor or any sub-contractor save and except to the extent that such accident or injury results from or is contributed to by any act or default of the Employer his agents or servants and the Contractor shall indemnify and keep indemnified the Employer against all such damages and compensation (save and except as aforesaid) and against all claims demands proceedings costs charges and expenses whatsoever in respect thereof or in relation thereto.

Remedy on Contractor's Failure to Insure

25. If the Contractor shall fail upon request to produce to the Employer satisfactory evidence that there is in force the insurance referred to in Clauses 21 and 23 or any other insurance which he may be required to effect under the terms of the Contract then and in any such case the Employer may effect and keep in force any such insurance and pay such premium or premiums as may be necessary for that purpose and from time to time deduct the amount so paid by the Employer as aforesaid from any monies due or which may become due to the Contractor or recover the same as a debt due from the Contractor.

Giving of Notices and Payment of Fees

26. (1) The Contractor shall save as provided in Clause 27 give all notices and pay all fees required to be given or paid by any Act of Parliament or any Regulation or Bye-law of any local or other statutory authority in relation to the execution of the Investigation and by the rules and regulations of all public bodies and companies whose property or rights are or may be affected in any way by the Investigation. The Employer shall repay or allow to the Contractor all such sums as the Engineer shall certify to have been properly payable and paid by the Contractor in respect of such fees and also all rates and taxes paid by the Contractor in respect of the Site or any part thereof or anything constructed or erected thereon or on any part thereof or any temporary structures situate elsewhere but used exclusively for the purposes of the Investigation or any structures used temporarily and exclusively for the purposes of the Investigation.

Contractor to Conform with Statutes, etc.

(2) The Contractor shall ascertain and conform in all respects with the provisions of any general or local Act of Parliament and the Regulations and Bye-laws of any local or other statutory authority which may be applicable to the Investigation and with such rules and regulations of public bodies and companies as aforesaid and shall keep the Employer indemnified against all penalties and liability of every kind for breach of any such Act Regulation or Bye-law. Provided always that:

(a) the Contractor shall not be required to indemnify the Employer against the consequences of any such breach which is the unavoidable result of complying with the Drawings Specification or instructions of the Engineer;

(b) if the Drawings Specification or instructions of the Engineer shall at any time be found not to be in conformity with any such Act Regulation or Bye-law the Engineer shall issue such instructions including the ordering of a variation under Clause 51 as may be necessary to ensure conformity with such Act Regulation or Bye-law;

(c) the Contractor shall not be responsible for obtaining any planning permission which may be necessary in respect of any Ancillary Works specified or designed by the Engineer and the Employer hereby warrants that all the said permissions have been or will in due time be obtained.

Public Utilities Street Works Act 1950—Definitions

27. (1) For the purposes of this Clause:

(a) the expression 'the Act' shall mean and include the Public Utilities Street Works Act 1950 and any statutory modification or re-enactment thereof for the time being in force;

(b) all other expressions common to the Act and to this Clause shall have the same meaning as that assigned to them by the Act.

Notifications by Employer to Contractor

(2) The Employer shall before the commencement of the Investigation notify the Contractor in writing:

(a) whether the Investigation or any parts thereof (and if so which parts) are Emergency Works; and

(b) which (if any) parts of the Investigation are to be carried out in Controlled Land or in a Prospectively Maintainable Highway.

If any duly authorised variation of the Investigation shall involve the execution of Site Operations in a Street or in Controlled Land or in a Prospectively Maintainable Highway or are Emergency Works the Employer shall notify the Contractor in writing accordingly at the time such variation is ordered.

(3) The Employer shall (subject to the obligations of the Contractor under sub-clause (4) of this Clause) serve all such notices as may from time to time whether before or during the course of or after completion of the Site Operations be required to be served under the Act.
Service of Notices by Employer

(4) The Contractor shall in relation to any part of the Investigation (other than Emergency Works) and subject to the compliance by the Employer with sub-clause (2) of this Clause give not less than 21 days' notice in writing to the Employer before:
Notices by Contractor to Employer

(a) commencing any part of the Investigation in a Street (as defined by Sections 1(3) and 38(1) of the Act); or

(b) commencing any part of the Investigation in Controlled Land or in a Prospectively Maintainable Highway; or

(c) commencing in a Street or in Controlled Land or in a Prospectively Maintainable Highway any part of the Investigation which is likely to affect the apparatus of any Owning Undertaker (within the meaning of Section 26 of the Act).

Such notice shall state the date on which and the place at which the Contractor intends to commence the execution of the work referred to therein.

(5) If the Contractor having given any such notice as is required by sub-clause (4) of this Clause shall not commence the part of the Investigation to which such notice relates within 2 months after the date when such notice is given such notice shall be treated as invalid and compliance with the said sub-clause (4) shall be requisite as if such notice had not been given.
Failure to Commence Street Works

(6) In the event of such a variation of the Investigation as is referred to in sub-clause (2) of this Clause being ordered by or on behalf of the Employer and resulting in delay in the execution of the Investigation by reason of the necessity of compliance by the Contractor with sub-clause (4) of this Clause the Engineer shall take such delay into account in determining any extension of time to which the Contractor is entitled under Clause 44 and the Contractor shall subject to Clause 52 be paid in accordance with Clause 60 such additional cost as the Engineer shall consider to have been reasonably attributable to such delay.
Delays Attributable to Variations

(7) Except as otherwise provided by this Clause where in relation to the carrying out of the Investigation the Act imposes any requirements or obligations upon the Employer the Contractor shall subject to Clause 49(5) comply with such requirements and obligations and shall (subject as aforesaid) indemnify the Employer against any liability which the Employer may incur in consequence of any failure to comply with the said requirements and obligations.
Contractor to Comply with Other Obligations of Act

28. (1) The Contractor shall save harmless and indemnify the Employer from and against all claims and proceedings for or on account of infringement of any patent rights design trade-mark or name or other protected rights in respect of any Equipment machine work or material used for or in connection with the Investigation and from and against all claims demands proceedings damages costs charges and expenses whatsoever in respect thereof or in relation thereto.
Patent Rights

(2) Except where otherwise specified the Contractor shall pay all tonnage and other royalties rent and other payments or compensation (if any) for getting stone sand gravel clay or other materials required for the Investigation.
Royalties

29. (1) All operations necessary for the execution of the Site Operations shall so far as compliance with the requirements of the Contract permits be carried on so as not to interfere unnecessarily or improperly with the public convenience or the access to or use or occupation of public or private roads and footpaths or to or of properties whether in the possession of the Employer or of any other person and the Contractor shall save harmless and indemnify the Employer in respect of all claims demands proceedings damages costs charges and expenses whatsoever arising out of or in relation to any such matters.
Interference with Traffic and Adjoining Properties

(2) All Site Operations shall be carried out without unreasonable noise and disturbance. The Contractor shall indemnify the Employer from and against any liability for damages on account of noise or other disturbance created while or in carrying out the Site Operations and from and against all claims demands proceedings damages costs charges and expenses whatsoever in regard or in relation to such liability.
Noise and Disturbance

30. (1) The Contractor shall use every reasonable means to prevent any of the highways or bridges communicating with or on the routes to the Site from being subjected to extraordinary traf-
Avoidance of Damage to Highways, etc.

135

fic within the meaning of the Highways Act 1959 or in Scotland the Road Traffic Act 1930 or any statutory modification or re-enactment thereof by any traffic of the Contractor or any of his sub-contractors and in particular shall select routes and use vehicles and restrict and distribute loads so that any such extraordinary traffic as will inevitably arise from the moving of Equipment and materials or manufactured or fabricated articles from and to the Site shall be limited as far as reasonably possible and so that no unnecessary damage or injury may be occasioned to such highways and bridges.

Transport of Equipment

(2)   Save insofar as the Contract otherwise provides the Contractor shall be responsible for and shall pay the cost of strengthening any bridges or altering or improving any highway communicating with the Site to facilitate the movement of Equipment required in the execution of the Investigation and the Contractor shall indemnify and keep indemnified the Employer against all claims for damage to any highway or bridge communicating with the Site caused by such movement including such claims as may be made by any competent authority directly against the Employer pursuant to any Act of Parliament or other Statutory Instrument and shall negotiate and pay all claims arising solely out of such damage.

Transport of Materials

(3)   If notwithstanding sub-clause (1) of this Clause any damage shall occur to any bridge or highway communicating with the Site arising from the transport of materials or manufactured or fabricated articles in the execution of the Investigation the Contractor shall notify the Engineer as soon as he becomes aware of such damage or as soon as he receives any claim from the authority entitled to make such claim. Where under any Act of Parliament or other Statutory Instrument the haulier of such materials or manufactured or fabricated articles is required to indemnify the highway authority against damage the Employer shall not be liable for any costs charges or expenses in respect thereof or in relation thereto. In other cases the Employer shall negotiate the settlement of and pay all sums due in respect of such claim and shall indemnify the Contractor in respect thereof and in respect of all claims demands proceedings damages costs charges and expenses in relation thereto. Provided always that if and so far as any such claim or part thereof shall in the opinion of the Engineer be due to any failure on the part of the Contractor to observe and perform his obligations under sub-clause (1) of this Clause then the amount certified by the Engineer to be due to such failure shall be paid by the Contractor to the Employer or deducted from any sum or which may become due to the Contractor.

Facilities for Other Contractors

31.   (1)   The Contractor shall in accordance with the requirements of the Engineer afford all reasonable facilities for any other contractors employed by the Employer and their workmen and for the workmen of the Employer and of any other properly authorised authorities or statutory bodies who may be employed in the execution on or near the Site of any work not in the Contract or of any contract which the Employer may enter into in connection with or ancillary to the Investigation.

Delay and Extra Cost

(2)   If compliance with sub-clause (1) of this Clause shall involve the Contractor in delay or cost beyond that reasonably to be foreseen by an experienced contractor at the time of tender then the Engineer shall take such delay into account in determining any extension of time to which the Contractor is entitled under Clause 44 and the Contractor shall subject to Clause 52(4) be paid in accordance with Clause 60 the amount of such cost as may be reasonable.

Fossils, etc.

32.   Without in any way limiting the execution of the Investigation all fossils coins articles of value or antiquity and structures or other remains or things of geological or archaeological interest discovered on the Site shall as between the Employer and the Contractor be deemed to be the absolute property of the Employer and the Contractor shall take reasonable precautions to prevent his workmen or any other persons from improperly removing or damaging any such article or thing and shall immediately upon discovery thereof and before removal acquaint the Engineer of such discovery and carry out at the expense of the Employer the Engineer's orders as to the disposal of the same.

Clearance of Site on Completion

33.   On the completion of the Site Operations the Contractor shall clear away and remove from the Site all Equipment surplus material and rubbish of every kind and leave the whole of the Site and any Ancillary Works clean and in a workmanlike condition to the satisfaction of the Engineer.

LABOUR

Rates of Wages/Hours and Conditions of Labour

34.   The Contractor shall in the execution of the Contract where appropriate pay rates of wages and observe the hours and conditions for the employment of operatives not less favourable than those established for the time being in the Working Rule Agreement of the Civil Engineering Construction Conciliation Board for Great Britain.

Returns of Labour and Equipment

35.   The Contractor shall if required by the Engineer deliver to the Engineer or at his office a return in such form and at such intervals as the Engineer may prescribe showing in detail the numbers of the several classes of labour from time to time employed by the Contractor on the Site and such information respecting Equipment as the Engineer may require. The Contractor shall require his sub-contractors to observe the provisions of this Clause.

WORKMANSHIP AND MATERIALS

36. (1) All materials supplied by the Contractor and workmanship shall be of the respective kinds described in the Contract and in accordance with the Engineer's instructions and shall be subjected from time to time to such tests as the Engineer may direct at the place of manufacture or fabrication or on the Site or such other places as may be specified in the Contract. The Contractor shall provide such assistance instruments machines labour and materials as are normally required for examining measuring and testing any Ancillary Works and the quality weight or quantity of any materials used and shall supply for testing samples of materials before incorporation in the Ancillary Works or use in the Site Operations as may be selected and required by the Engineer. **Quality of Materials and Workmanship and Tests**

(2) All samples of materials for Ancillary Works shall be supplied by the Contractor at his own cost if the supply thereof is clearly intended by or provided for in the Contract but if not then at the cost of the Employer. **Cost of Samples**

(3) All Equipment shall be of the respective type and kind suitable for the proper execution of the Investigation in accordance with the Engineer's instructions and shall be calibrated and subjected to such calibration tests as may be necessary to ensure performance in accordance with the requirements of the Contract. **Suitability and Calibration of Equipment**

(4) The cost of making any test or calibration pursuant to this clause shall be borne by the Contractor if such tests or calibrations are clearly intended by or provided for in the Contract and in the case of a test to ascertain whether the design of any finished or partially finished work is appropriate for the purposes which it was intended to fulfil if it is particularised in the Specification or Bill of Quantities in sufficient detail to enable the Contractor to have priced or allowed for the same in his Tender. If any test or calibration is ordered by the Engineer which is either not so intended by or provided for or is not so particularised then the cost of such test or calibration shall be borne by the Contractor if the test or calibration shows the Ancillary Works Equipment workmanship or materials not to be in accordance with the provisions of the Contract or the Engineer's instructions but otherwise by the Employer. **Cost of Tests or Calibrations**

37. The Engineer and any person authorised by him shall at all times have access to the Site and to the Site Operations and Laboratory Testing and to samples wherever stored and to all laboratories workshops and places where work is being prepared or carried out or whence materials manufactured articles and machinery are being obtained for the Investigation and the Contractor shall afford every facility for and every assistance in or in obtaining the right to such access. **Access by Engineer**

38. (1) No trial pits or Ancillary Works shall be covered up or put out of view without the approval of the Engineer and the Contractor shall afford full opportunity for the Engineer to examine and measure any trial pits or Ancillary Works which are about to be covered up or put out of view. The Contractor shall give due notice to the Engineer whenever any such trial pits or Ancillary Works are ready or about to be ready for examination and the Engineer shall without unreasonable delay unless he considers it unnecessary and advises the Contractor accordingly attend for the purpose of examining and measuring such trial pits or Ancillary Works. **Examination of Work before Covering up**

(2) The Contractor shall uncover any part or parts of the trial pits or Ancillary Works as the Engineer may from time to time direct and shall reinstate and make good such part or parts to the satisfaction of the Engineer. If any such part or parts have been covered up or put out of view after compliance with the requirements of sub-clause (1) of this Clause and are found to be executed in accordance with the Contract the cost of uncovering reinstating and making good the same shall be borne by the Employer but in any other case all such cost shall be borne by the Contractor. **Uncovering and Making Openings**

39. (1) The Engineer shall during the progress of the Site Operations have power to order in writing: **Removal of Improper Work and Materials**

- (a) the removal from the Site within such time or times as may be specified in the order of any materials which in the opinion of the Engineer are not in accordance with the Contract;
- (b) the substitution of proper and suitable materials; and
- (c) the rectification or the removal and proper re-execution (notwithstanding any previous test thereof or interim payment therefor) of any work which in respect of materials or workmanship is not in the opinion of the Engineer in accordance with the Contract.

(2) In case of default on the part of the Contractor in carrying out such order the Employer shall be entitled to employ and pay other persons to carry out the same and all expenses consequent thereon or incidental thereto shall be borne by the Contractor and shall be recoverable from him by the Employer or may be deducted by the Employer from any monies due or which may become due to the Contractor. **Default of Contractor in Compliance**

(3) Failure of the Engineer or any person acting under him pursuant to Clause 2 to dis- **Failure to Disapprove**

approve any work or materials shall not prejudice the power of the Engineer or any of them subsequently to disapprove such work or materials.

Suspension of Investigation

40. (1)　The Contractor shall on the written order of the Engineer suspend the progress of the Investigation or any part thereof for such time or times and in such manner as the Engineer may consider necessary and shall during such suspension properly protect and secure the work so far as it is necessary in the opinion of the Engineer. Subject to Clause 52(4) the Contractor shall be paid in accordance with Clause 60 the extra cost (if any) incurred in giving effect to the Engineer's instructions under this Clause except to the extent that such suspension is:

(a)　otherwise provided for in the Contract; or

(b)　necessary by reason of weather conditions or by some default on the part of the Contractor; or

(c)　necessary for the proper execution of the work or for the safety of the Site Operations or any part thereof inasmuch as such necessity does not arise from any act or default of the Engineer or the Employer or from any of the Excepted Risks defined in Clause 20.

The Engineer shall take any delay occasioned by a suspension ordered under this Clause (including that arising from any act or default of the Engineer or the Employer) into account in determining any extension of time to which the Contractor is entitled under Clause 44 except when such suspension is otherwise provided for in the Contract or is necessary by reason of some default on the part of the Contractor.

Suspension lasting more than Three Months

(2)　If the progress of the Investigation or any part thereof is suspended on the written order of the Engineer and if permission to resume work is not given by the Engineer within a period of 3 months from the date of suspension then the Contractor may unless such suspension is otherwise provided for in the Contract or continues to be necessary by reason of some default on the part of the Contractor serve a written notice on the Engineer requiring permission within 28 days from the receipt of such notice to proceed with the Investigation or that part thereof in regard to which progress is suspended. If within the said 28 days the Engineer does not grant such permission the Contractor by a further written notice so served may (but is not bound to) elect to treat the suspension where it affects part only of the Investigation as an omission of such part under Clause 51 or where it affects the whole of the Investigation as an abandonment of the Contract by the Employer.

COMMENCEMENT TIME AND DELAYS

Commencement of Investigation

41.　The Contractor shall commence the Investigation on or as soon as is reasonably possible after the Date for Commencement of the Investigation to be notified by the Engineer in writing which date shall be within a reasonable time after the date of acceptance of the Tender. Thereafter the Contractor shall proceed with the Investigation with due expedition and without delay in accordance with the Contract.

Possession of Site

42. (1)　Save insofar as the Contract may prescribe the extent of portions of the Site of which the Contractor is to be given possession from time to time and the order in which such portions shall be made available to him and subject to any requirement in the Contract as to the order in which the Investigation shall be executed the Employer will at the Date for Commencement of the Investigation notified under Clause 41 give to the Contractor possession of so much of the Site as may be required to enable the Contractor to commence and proceed with the Investigation in accordance with the programme referred to in Clause 14 and will from time to time as the Investigation proceeds give to the Contractor possession of such further portions of the Site as may be required to enable the Contractor to proceed with the Investigation with due despatch in accordance with the said programme. If the Contractor suffers delay or incurs cost from failure on the part of the Employer to give possession in accordance with the terms of this Clause then the Engineer shall take such delay into account in determining any extension of time to which the Contractor is entitled under Clause 44 and the Contractor shall subject to Clause 52(4) be paid in accordance with Clause 60 the amount of such cost as may be reasonable.

Wayleaves, etc.

(2)　The Employer shall bear all expenses and charges and make payments for special or temporary wayleaves required for access to the Site for the purposes of the Investigation. The Contractor shall provide at his own cost any additional accommodation outside the Site other than necessary access required by him for the purposes of the Investigation.

Time for Completion

43.　The whole of the Investigation and any Section required to be completed within a particular time as stated in the Appendix to the Form of Tender shall be completed within the time so stated (or such extended time as may be allowed under Clause 44) calculated from the Date for Commencement of the Investigation notified under Clause 41.

Extension of Time for Completion

44. (1)　Should any variation ordered under Clause 51(1) or increased quantities referred to in Clause 51(3) or any other cause of delay referred to in these Conditions or exceptional adverse weather conditions or other special circumstances of any kind whatsoever which may occur be such as fairly to entitle the Contractor to an extension of time for the completion of the Investigation or (where different periods for completion of different Sections are provided for in the Appendix to the Form of Tender) of the relevant Section the Contractor shall within 28 days after the cause of the delay has arisen or as soon thereafter as is reasonable in all the circumstances deliver to the Engineer full and detailed particulars of any claim to extension of time to which he may consider himself entitled in order that such claim may be investigated at the time.

    (2)   The Engineer shall upon receipt of such particulars or if he thinks fit in the absence of any such claim consider all the circumstances known to him at that time and make an assessment of the extension of time (if any) to which he considers the Contractor entitled for the completion of the Investigation or relevant Section and shall by notice in writing to the Contractor grant such extension of time for completion. In the event that the Contractor shall have made a claim for an extension of time but the Engineer considers the Contractor not entitled thereto the Engineer shall so inform the Contractor.   **Interim Assessment of Extension**

    (3)   The Engineer shall at or as soon as possible after the due date or extended date for completion (and whether or not the Contractor shall have made any claim for an extension of time) consider all the circumstances known to him at that time and take action similar to that provided for in sub-clause (2) of this Clause. Should the Engineer consider that the Contractor is not entitled to an extension of time he shall so notify the Employer and the Contractor.   **Assessment at Due Date for Completion**

    (4)   The Engineer shall upon the issue of the Certificate of Completion of the Investigation or of the relevant Section review all the circumstances of the kind referred to in sub-clause (1) of this Clause and shall finally determine and certify to the Contractor the overall extension of time (if any) to which he considers the Contractor entitled in respect of the Investigation or any relevant Section. No such final review of the circumstances shall result in a decrease in any extension of time already granted by the Engineer pursuant to sub-clauses (2) or (3) of this Clause.   **Final Determination of Extension**

45.   Subject to any provision to the contrary contained in the Contract none of the Site Operations shall be executed during the night or on Sundays without the permission in writing of the Engineer save when the work is unavoidable or absolutely necessary for the saving of life or property or for the safety of the Site Operations in which case the Contractor shall immediately advise the Engineer or the Engineer's Representative. Provided always that this Clause shall not be applicable in the case of any work which it is customary to carry out outside normal working hours or by rotary or double shifts.   **Night and Sunday Work**

46.   If for any reason which does not entitle the Contractor to an extension of time the rate of progress of the Investigation or any Section is at any time in the opinion of the Engineer too slow to ensure completion by the prescribed time or extended time for completion the Engineer shall so notify the Contractor in writing and the Contractor shall thereupon take such steps as are necessary and the Engineer may approve to expedite progress so as to complete the Investigation or such Section by the prescribed time or extended time. The Contractor shall not be entitled to any additional payment for taking such steps. If as a result of any notice given by the Engineer under this Clause the Contractor shall seek the Engineer's permission to do any work at night or on Sundays such permission shall not be unreasonably refused.   **Rate of Progress**

LIQUIDATED DAMAGES AND LIMITATION OF DAMAGES FOR DELAYED COMPLETION

47. (1) (a)   In the Appendix to the Form of Tender under the heading 'Liquidated Damages for Delay' there is stated in column 1 the sum which represents the Employer's genuine pre-estimate (expressed as a rate per week or per day as the case may be) of the damages likely to be suffered by him in the event that the whole of the Investigation shall not be completed within the time prescribed by Clause 43.   **Liquidated Damages for Whole of Investigation**

    Provided that in lieu of such sum there may be stated such lesser sum as represents the limit of the Contractor's liability for damages for failure to complete the whole of the Investigation within the time for completion therefor or any extension thereof granted under Clause 44.

    (b)   If the Contractor should fail to complete the whole of the Investigation within the prescribed time or any extension thereof granted under Clause 44 the Contractor shall pay to the Employer for such default the sum stated in column 1 aforesaid for every week or day as the case may be which shall elapse between the date on which the prescribed time or any extension thereof expired and date of completion of the whole of the Investigation. Provided that if any part of the Investigation not being a Section or part of a Section shall be certified as complete pursuant to Clause 48 before completion of the whole of the Investigation the sum stated in column 1 shall be reduced by the proportion which the value of the part completed bears to the value of the whole of the Investigation.

  (2) (a)   In cases where any Section shall be required to be completed within a particular time as stated in the Appendix to the Form of Tender there shall also be stated in the   **Liquidated Damages for Sections**

said Appendix under the heading 'Liquidated Damages for Delay' in column 2 the sum by which the damages stated in column 1 or the limit of the Contractor's said liability as the case may be shall be reduced upon completion of each such Section and in column 3 the sum which represents the Employer's genuine pre-estimate (expressed as aforesaid) of any specific damage likely to be suffered by him in the event that such Section shall not be completed within that time.

Provided that there may be stated in column 3 in lieu of such sum such lesser sum as represents the limit of the Contractor's liability for failure to complete the relevant Section within the relevant time.

(*b*) If the Contractor should fail to complete any Section within the relevant time for completion or any extension thereof granted under Clause 44 the Contractor shall pay to the Employer for such default the sum stated in column 3 aforesaid for every week or day as the case may be which shall elapse between the date on which the relevant time or any extension thereof expired and the date of completion of the relevant Section. Provided that:

(i) if completion of a Section shall be delayed beyond the due date for completion of the whole of the Investigation the damages payable under sub-clauses (1) and (2) of this Clause until completion of that Section shall be the sum stated in column 1 plus in respect of that Section the sum stated in column 3 less the sum stated in column 2;

(ii) if any part of a Section shall be certified as complete pursuant to Clause 48 before completion of the whole thereof the sums stated in columns 2 and 3 in respect of that Section shall be reduced by the proportion which the value of the part bears to the value of the Section and the sums stated in column 1 shall be reduced by the same amount as the sum in column 2 is reduced; and

(iii) upon completion of any such Section the sum stated in column 1 shall be reduced by the sum stated in column 2 in respect of that Section at the date of such completion.

Damages not a Penalty

(3) All sums payable by the Contractor to the Employer pursuant to this Clause shall be paid as liquidated damages for delay and not as a penalty.

Deduction of Liquidated Damages

(4) If the Engineer shall under Clause 44(3) or (4) have determined and certified any extension of time to which he considers the Contractor entitled or shall have notified the Employer and the Contractor that he is of the opinion that the Contractor is not entitled to any or any further extension of time the Employer may deduct and retain from any sum otherwise payable by the Employer to the Contractor hereunder the amount which in the event that the Engineer's said opinion should not be subsequently revised would be the amount of the liquidated damages payable by the Contractor under this Clause.

Reimbursement of Liquidated Damages

(5) If upon a subsequent or final review of the circumstances causing delay the Engineer shall grant an extension or further extension of time or if an arbitrator appointed under Clause 66 shall decide that the Engineer should have granted such an extension or further extension of time the Employer shall no longer be entitled to liquidated damages in respect of the period of such extension of time. Any sums in respect of such period which may have been recovered pursuant to sub-clause (3) of this Clause shall be reimbursable forthwith to the Contractor together with interest at the rate provided for in Clause 60(6) from the date on which such liquidated damages were recovered from the Contractor.

COMPLETION CERTIFICATE

Certificate of Completion

48. (1) When the Contractor shall consider that the whole of the Investigation has been substantially completed in accordance with the requirements of the Contract he may give notice to that effect to the Engineer or to the Engineer's Representative. In all cases such notice shall be in writing and shall be deemed to be a request by the Contractor for the Engineer to issue a Certificate of Completion in respect of the Investigation and the Engineer shall within 14 days of the delivery of such notice either issue to the Contractor (with a copy to the Employer) a Certificate of Completion stating the date on which in his opinion the Investigation was substantially completed in accordance with the Contract or else give instructions in writing to the Contractor specifying all the work which in the Engineer's opinion requires to be done before the issue of such Certificate. If the Engineer shall give such instructions the Contractor shall be entitled to receive such Certificate of Completion within 21 days of completion to the satisfaction of the Engineer of the work specified in the said instructions.

Completion of Sections

(2) (*a*) Similarly in accordance with the procedure set out in sub-clause (1) of this Clause the Contractor may request and the Engineer shall issue a Certificate of Completion in respect of:

(i) the whole of the Site Operations (which for the purpose of Clause 47 shall be deemed to be a Section of the Investigation);

140

(ii) any Section in respect of which a separate time for completion is provided in the Appendix to the Form of Tender;

(iii) any substantial part of the Investigation which has been completed to the satisfaction of the Engineer and is occupied or used by the Employer.

(b) Where a Contractor requests a Certificate of Completion for the Site Operations or where any Section comprises Site Operations the notice in sub-clause (1) of this Clause shall be accompanied by an undertaking in writing by the Contractor to finish any outstanding work during the Period of Maintenance.

(3)   If the Engineer shall be of the opinion that any part of the Investigation shall have been substantially completed in accordance with the Contract he may issue a Certificate of Completion in respect of that part of the Investigation before completion of the whole of the Investigation. Where such part shall comprise Site Operations upon the issue of such certificate the Contractor shall be deemed to have undertaken to complete any outstanding work in that part of the Site Operations during the Period of Maintenance.

Completion of Other Parts of Site Operations

(4)   Provided always that a Certificate of Completion given in respect of any Section or part of the Investigation before completion of the whole shall not be deemed to certify completion of any ground or surfaces requiring reinstatement unless such certificate shall expressly so state.

Reinstatement of Ground

MAINTENANCE AND DEFECTS

49.   (1)   In these Conditions the expression 'Period of Maintenance' shall mean the period of maintenance named in the Appendix to the Form of Tender calculated from the date of completion of the Site Operations or any Section or part thereof certified by the Engineer in accordance with Clause 48 as the case may be.

Definition of 'Period of Maintenance'

(2)   To the intent that the Site and the Ancillary Works and any Section or part thereof shall at or as soon as practicable after the expiration of the relevant Period of Maintenance be delivered up to the Employer in the condition required by the Contract (fair wear and tear excepted) to the satisfaction of the Engineer the Contractor shall finish the work (if any) outstanding at the date of completion as certified under Clause 48 as soon as may be practicable after such date and shall execute all such work of repair amendment reconstruction rectification and making good of defects imperfections shrinkages damage subsidence of backfill or other faults resulting from the execution of the Site Operations as may during the Period of Maintenance or within 14 days after its expiration be required of the Contractor in writing by the Engineer as a result of an inspection made by or on behalf of the Engineer prior to its expiration.

Execution of Work of Repair, etc.

(3)   All such work shall be carried out by the Contractor at his own expense if the necessity thereof shall in the opinion of the Engineer be due to the use of materials or workmanship not in accordance with the Contract or to neglect or failure on the part of the Contractor to comply with any obligation expressed or implied on the Contractor's part under the Contract. If in the opinion of the Engineer such necessity shall be due to any other cause the value of such work shall be ascertained and paid for as if it were additional work.

Cost of Execution of Work of Repair, etc.

(4)   If the Contractor shall fail to do any such work as aforesaid required by the Engineer the Employer shall be entitled to carry out such work by his own workmen or by other contractors and if such work is work which the Contractor should have carried out at the Contractor's own cost the Employer shall be entitled to recover from the Contractor the cost thereof or may deduct the same from any monies due or that become due to the Contractor.

Remedy on Contractor's Failure to carry out Work Required

(5)   Provided always that if in the course or for the purposes of the execution of the Site Operations or any part thereof any highway or other road or way shall have been broken into then notwithstanding anything herein contained:

Temporary Reinstatement

(a) If the permanent reinstatement of such highway or other road or way is to be carried out by the appropriate Highway Authority or by some person other than the Contractor (or any sub-contractor to him) the Contractor shall at his own cost and independently of any requirements of or notice from the Engineer be responsible for the making good of any subsidence or shrinkage or other defect imperfection or fault in the temporary reinstatement of such highway or other road or way and for the execution of any necessary repair or amendment thereof from whatever cause the necessity arises until the end of the Period of Maintenance in respect of the works beneath such highway or other road or way or until the Highway Authority or other person as aforesaid shall have taken possession of the Site for the purpose of carrying out permanent reinstatement (whichever is the earlier) and shall indemnify and save harmless the Employer against and from any damage or injury to the Employer or to third parties arising out or in consequence of any neglect or failure of the Contractor to comply with the foregoing obligations or any of them and against and from all claims demands proceedings damages costs charges and expenses whatsoever in respect thereof or in relation thereto. As from the end of such

141

Period of Maintenance or the taking possession as aforesaid (whichever shall first happen) the Employer shall indemnify and save harmless the Contractor against and from any damage or injury as aforesaid arising out or in consequence of or in connection with the said permanent reinstatement or any defect imperfection or failure of or in such work of permanent reinstatement and against and from all claims demands proceedings damages costs charges and expenses whatsoever in respect thereof or in relation thereto.

(b) Where the Highway Authority or other person as aforesaid shall take possession of the Site as aforesaid in sections or lengths the responsibility of the Contractor under paragraph (a) of this sub-clause shall cease in regard to any such section or length at the time possession thereof is so taken but shall during the continuance of the said Period of Maintenance continue in regard to any length of which possession has not been so taken and the indemnities given by the Contractor and the Employer respectively under the said paragraph shall be construed and have effect accordingly.

Contractor to Search

50. The Contractor shall if required by the Engineer in writing carry out such searches tests or trials as may be necessary to determine the cause of any defect imperfection or fault in any Ancillary Works under the directions of the Engineer. Unless such defect imperfection or fault shall be one for which the Contractor is liable under the Contract the cost of the work carried out by the Contractor as aforesaid shall be borne by the Employer. But if such defect imperfection or fault in any Ancillary Works shall be one for which the Contractor is liable the cost of the work carried out as aforesaid shall be borne by the Contractor and he shall in such case repair rectify and make good such defect imperfection or fault at his own expense in accordance with Clause 49.

ALTERATIONS ADDITIONS AND OMISSIONS

Ordered Variations

51. (1) The Engineer shall order any variation to any part of the Investigation that may in his opinion be necessary for the completion of the Investigation and shall have power to order any variation that for any other reason shall in his opinion be desirable for the satisfactory completion of the Investigation. Such variations may include additions omissions substitutions alterations changes in quality form character kind position dimension level or line and changes in the specified sequence method or timing of the Investigation if any.

Ordered Variations to be in Writing

(2) Other than as provided for in sub-clause (4) of Clause 13 no such variation shall be made by the Contractor without an order by the Engineer. All such orders shall be given in writing provided that if for any reason the Engineer shall find it necessary to give any such order orally in the first instance the Contractor shall comply with such oral order. Such oral order shall be confirmed in writing by the Engineer as soon as is possible in the circumstances. If the Contractor shall confirm in writing to the Engineer any oral order by the Engineer and such confirmation shall not be contradicted in writing by the Engineer forthwith it shall be deemed to be an order in writing by the Engineer. No variation ordered or deemed to be ordered in writing in accordance with sub-clauses (1) and (2) of this Clause shall in any way vitiate or invalidate the Contract but the value (if any) of all such variations shall be taken into account in ascertaining the amount of the Contract Price.

Changes in Quantities

(3) No order in writing shall be required for increase or decrease in the quantity of any work where such increase or decrease is not the result of an order given under this Clause but is the result of the quantities exceeding or being less than those stated in the Bill of Quantities.

Valuation of Ordered Variations

52. (1) The value of all variations ordered by the Engineer in accordance with Clause 51 shall be ascertained by the Engineer after consultation with the Contractor in accordance with the following principles. Where work is of similar character and executed under similar conditions to work priced in the Bill of Quantities it shall be valued at such rates and prices contained therein as may be applicable. Where work is not of a similar character or is not executed under similar conditions the rates and prices in the Bill of Quantities shall be used as the basis for valuation so far as may be reasonable failing which a fair valuation shall be made. Failing agreement between the Engineer and the Contractor as to any rate or price to be applied in the valuation of any variation the Engineer shall determine the rate or price in accordance with the foregoing principles and he shall notify the Contractor accordingly.

Engineer to Fix Rates

(2) Provided that if the nature or amount of any variation relative to the nature or amount of the whole of the contract work or to any part thereof shall be such that in the opinion of the Engineer or the Contractor any rate or price contained in the Contract for any item of work is by reason of such variation rendered unreasonable or inapplicable either the Engineer shall give to the Contractor or the Contractor shall give to the Engineer notice before the varied work is commenced or as soon thereafter as is reasonable in all the circumstances that such rate or price should be varied and the Engineer shall fix such rate or price as in the circumstances he shall think reasonable and proper.

Daywork

(3) The Engineer may if in his opinion it is necessary or desirable order in writing that any additional or substituted work shall be executed on a daywork basis. The Contractor shall then be

paid for such work under the conditions set out in the Site Operations Daywork Schedule included in the Bill of Quantities and at the rates and prices affixed thereto by him in his Tender. In the absence of any particular rate or price the Engineer shall fix such rate or price as in the circumstances he shall think reasonable and proper.

The Contractor shall furnish to the Engineer such receipts or other vouchers as may be necessary to prove the amounts paid and before ordering materials shall submit to the Engineer quotations for the same for his approval.

In respect of all work executed on a daywork basis the Contractor shall during the continuance of such work deliver each day to the Engineer's Representative an exact list in duplicate of the names occupation and time of all workmen employed on such work and a statement also in duplicate showing the description and quantity of all materials and plant used thereon or therefor (other than plant which is included in the percentge addition in accordance with the Site Operations Daywork Schedule under which payment for daywork is made). One copy of each list and statement will if correct or when agreed be signed by the Engineer's Representative and returned to the Contractor. At the end of each month the Contractor shall deliver to the Engineer's Representative a priced statement of the labour material and plant (except as aforesaid) used and the Contractor shall not be entitled to any payment unless such lists and statements have been fully and punctually rendered. Provided always that if the Engineer shall consider that for any reason the sending of such list or statement by the Contractor in accordance with the foregoing provision was impracticable he shall nevertheless be entitled to authorise payment for such work either as daywork (on being satisfied as to the time employed and plant and materials used on such work) or at such value therefor as he shall consider fair and reasonable.

(4) (a) If the Contractor intends to claim a higher rate or price than one notified to him by the Engineer pursuant to sub-clauses (1) and (2) of this Clause or Clause 56(2) the Contractor shall within 28 days after such notification give notice in writing of his intention to the Engineer. **Notice of Claims**

(b) If the Contractor intends to claim any additional payment pursuant to any Clause of these Conditions other than sub-clauses (1) and (2) of this Clause he shall give notice in writing of his intention to the Engineer as soon as reasonably possible after the happening of the events giving rise to the claim. Upon the happening of such events the Contractor shall keep such contemporary records as may reasonably be necessary to support any claim he may subsequently wish to make.

(c) Without necessarily admitting the Employer's liability the Engineer may upon receipt of a notice under this Clause instruct the Contractor to keep such contemporary records or further contemporary records as the case may be as are reasonable and may be material to the claim of which notice has been given and the Contractor shall keep such records. The Contractor shall permit the Engineer to inspect all records kept pursuant to this Clause and shall supply him with copies thereof as and when the Engineer shall so instruct.

(d) After the giving of a notice to the Engineer under this Clause the Contractor shall as soon as is reasonable in all the circumstances send to the Engineer a first interim account giving full and detailed particulars of the amount claimed to that date and of the grounds upon which the claim is based. Thereafter at such intervals as the Engineer may reasonably require the Contractor shall send to the Engineer further up to date accounts giving the accumulated total of the claim and any further grounds upon which it is based.

(e) If the Contractor fails to comply with any of the provisions of this Clause in respect of any claim which he shall seek to make then the Contractor shall be entitled to payment in respect thereof only to the extent that the Engineer has not been prevented from or substantially prejudiced by such failure in investigating the said claim.

(f) The Contractor shall be entitled to have included in any interim payment certified by the Engineer pursuant to Clause 60 such amount in respect of any claim as the Engineer may consider due to the Contractor provided that the Contractor shall have supplied sufficient particulars to enable the Engineer to determine the amount due. If such particulars are insufficient to substantiate the whole of the claim the Contractor shall be entitled to payment in respect of such part of the claim as the particulars may substantiate to the satisfaction of the Engineer.

PROPERTY IN MATERIALS AND PLANT

53. (1) For the purpose of this Clause: **Plant, etc.— Definitions**

(a) the expression 'Plant' shall mean any Equipment and any materials for Equipment but shall exclude any vehicles engaged in transporting any labour plant or materials to or from the Site;

(b) the expression 'agreement for hire' shall be deemed not to include an agreement for hire purchase.

Vesting of Plant

(2)　All Plant goods and materials owned by the Contractor or by any company in which the Contractor has a controlling interest shall when on the Site be deemed to be the property of the Employer.

Conditions of Hire of Plant

(3)　With a view to securing in the event of a forfeiture under Clause 63 the continued availability for the purpose of executing the Site Operations of any hired Plant the Contractor shall not bring on to the Site any hired Plant unless there s an agreement for the hire thereof which contains a provision that the owner thereof will on request in writing made by the Employer within 7 days after the date on which any forfeiture has become effective and on the Employer undertaking to pay all hire charges in respect thereof from such date hire such Plant to the Employer on the same terms in all respects as the same was hired to the Contractor save that the Employer shall be entitled to permit the use thereof by any other contractor employed by him for the purpose of completing the Site Operations under the terms of the said Clause 63.

Costs for Purposes of Clause 63

(4)　In the event of the Employer entering into any agreement for the hire of Plant pursuant to sub-clause (3) of this Clause all sums properly paid by the Employer under the provisions of any such agreement and all expenses incurred by him (including stamp duties) in entering into such agreement shall be deemed for the purpose of Clause 63 to be part of the cost of completing the Site Operations.

Notification of Plant Ownership

(5)　The Contractor shall upon request made by the Engineer at any time in relation to any item of Plant forthwith notify to the Engineer in writing the name and address of the owner thereof and shall in the case of hired Plant certify that the agreement for the hire thereof contains a provision in accordance with the requirement of sub-clause (3) of this Clause.

Irremovability of Plant, etc.

(6)　No Plant (except hired Plant) goods or materials or any part thereof shall be removed from the Site without the written consent of the Engineer which consent shall not be unreasonably withheld where the same are no longer immediately required for the purposes of the completion of the Site Operations but the Employer will permit the Contractor the exclusive use of all such Plant goods and materials in and for the completion of the Site Operations until the occurrence of any event which gives the Employer the right to exclude the Contractor from the Site and proceed with the completion of the Site Operations.

Revesting and Removal of Plant

(7)　Upon the removal of any such Plant goods or materials as have been deemed to have become the property of the Employer under sub-clause (2) of this Clause with the consent as aforesaid the property therein shall be deemed to revest in the Contractor and upon completion of the Site Operations the property in the remainder of such Plant goods and materials as aforesaid shall subject to Clause 63 be deemed to revest in the Contractor.

Disposal of Plant

(8)　If the Contractor shall fail to remove any Plant goods or materials as required pursuant to Clause 33 within such reasonable time after completion of the Site Operations as may be allowed by the Engineer then the Employer may:

(a)　sell any which are the property of the Contractor; and
(b)　return any not the property of the Contractor to the owner thereof at the Contractor's expense;

and after deducting from any proceeds of sale the costs charges and expenses of and in connection with such sale and of and in connection with return as aforesaid shall pay the balance (if any) to the Contractor but to the extent that the proceeds of any sale are insufficient to meet all such costs charges and expenses the excess shall be a debt due from the Contractor to the Employer and shall be deductible or recoverable by the Employer from any monies due or that may become due to the Contractor under the contract or may be recovered by the Employer from the Contractor at law.

Liability for Loss or Injury to Plant

(9)　The Employer shall not at any time be liable for the loss of or injury to any of the Plant goods or materials which have been deemed to become the property of the Employer under sub-clause (2) of this Clause save as mentioned in Clauses 20 and 65.

Incorporation of Clause in Sub-contracts

(10)　The Contractor shall where entering into any sub-contract for the execution of any part of the Site Operations incorporate in such sub-contract (by reference or otherwise) the provisions of this Clause in relation to Plant goods or materials brought on to the Site by the sub-contractor.

No Approval by Vesting

(11)　The operation of this Clause shall not be deemed to imply any approval by the Engineer of the materials or other matters referred to herein nor shall it prevent the rejection of any such materials at any time by the Engineer.

54.　Clause not used.

MEASUREMENT

Quantities

55.　(1)　The quantities set out in the Bill of Quantities are the estimated quantities of the work to be done in carrying out the Investigation but are not to be taken as the actual and correct quantities of such work to be executed by the Contractor in fulfilment of his obligations under the Contract.

(2)   Any error in description in the Bill of Quantities or omission therefrom shall not vitiate the Contract nor release the Contractor from the execution of the whole or any part of the Investigation according to the Drawings Specification and Schedules or from any of his obligations or liabilities under the Contract. Any such error or omission shall be corrected by the Engineer and the value of the work actually carried out shall be ascertained in accordance with Clause 52. Provided that there shall be no rectification of any errors omissions or wrong estimates in the descriptions rates and prices inserted by the Contractor in the Bill of Quantities.

Correction of Errors

56.   (1)   The Engineer shall except as otherwise stated ascertain and determine by admeasurement the value in accordance with the Contract of the work done in accordance with the Contract.

Measurement and Valuation

(2)   Should the actual quantities executed in respect of any item be greater or less than those stated in the Bill of Quantities and if in the opinion of the Engineer such increase or decrease of itself shall so warrant the Engineer shall after consultation with the Contractor determine an appropriate increase or decrease of any rates or prices rendered unreasonable or inapplicable in consequence thereof and shall notify the Contractor accordingly.

Increase or Decrease of Rate

(3)   The Engineer shall when he requires any part or parts of the work to be measured give reasonable notice to the Contractor who shall attend or send a qualified agent to assist the Engineer or the Engineer's Representative in making such measurement and shall furnish all particulars required by either of them. Should the Contractor not attend or neglect or omit to send such agent then the measurement made by the Engineer or approved by him shall be taken to be the correct measurement of the work.

Attending for Measurement

57.   Except where any statement or general or detailed description of the work in the Bill of Quantities expressly shows to the contrary Bills of Quantities shall be deemed to have been prepared and measurements shall be made according to the procedure set forth in the 'Civil Engineering Standard Method of Measurement' issued by the Institution of Civil Engineers in 1976 or such later or amended edition thereof as may be stated in the Appendix to the Form of Tender to have been adopted in its preparation notwithstanding any general or local custom.

Method of Measurement

PROVISIONAL AND PRIME COST SUMS AND NOMINATED SUB-CONTRACTS

58.   (1)   'Provisional Sum' means a sum included in the Contract and so designated for the execution of work or the supply of goods materials or services or for contingencies which sum may be used in whole or in part or not at all at the direction and discretion of the Engineer.

Provisional Sum

(2)   'Prime Cost (PC) Item' means an item in the Contract which contains (either wholly or in part) a sum referred to as Prime Cost (PC) which will be used for the execution of work or for the supply of goods materials or services for the Investigation.

Prime Cost Item

(3)   If in connection with any Provisional Sum or Prime Cost Item the services to be provided include any matter of design or specification of any part of the Ancillary Works or of any equipment or plant to be incorporated therein such requirement shall be expressly stated in the Contract and shall be included in any Nominated Sub-contract. The obligation of the Contractor in respect thereof shall be only that which has been expressly stated in accordance with this sub-clause.

Design Requirements to be Expressly Stated

(4)   In respect of every Prime Cost Item the Engineer shall have power to order the Contractor to employ a sub-contractor nominated by the Engineer for the execution of any work or the supply of any goods materials or services included therein. The Engineer shall also have power with the consent of the Contractor to order the Contractor to execute any such work or to supply any such goods materials or services in which event the Contractor shall be paid in accordance with the terms of a quotation submitted by him and accepted by the Engineer or in the absence thereof the value shall be determined in accordance with Clause 52.

Use of Prime Cost Items

(5)   All specialists merchants tradesmen and others nominated in the Contract for a Prime Cost Item or ordered by the Engineer to be employed by the Contractor in accordance with sub-clause (4) or sub-clause (7) of this Clause for the execution of any work or the supply of any goods materials or services are referred to in this Contract as 'Nominated Sub-contractors'.

Nominated Sub-contractors—Definition

(6)   The Contractor shall when required by the Engineer produce all quotations invoices vouchers sub-contract documents accounts and receipts in connection with expenditure in respect of work carried out by all Nominated Sub-contractors.

Production of Vouchers, etc.

(7)   In respect of every Provisional Sum the Engineer shall have power to order either or both of the following:

Use of Provisional Sums

> (a)   work to be executed or goods materials or services to be supplied by the Contractor the value of such work executed or goods materials or services supplied being determined in accordance with Clause 52 and included in the Contract Price;
> (b)   work to be executed or goods materials or services to be supplied by a Nominated Sub-contractor in accordance with Clause 59A.

Nominated Sub-contractors—Objection to Nomination

59A. (1)   Subject to sub-clause (2)(c) of this Clause the Contractor shall not be under any obligation to enter into any sub-contract with any Nominated Sub-contractor against whom the Contractor may raise reasonable objection or who shall decline to enter into a sub-contract with the Contractor containing provisions:

(*a*)   that in respect of the work goods materials or services the subject of the sub-contract the Nominated Sub-contractor will undertake towards the Contractor such obligations and liabilities as will enable the Contractor to discharge his own obligations and liabilities towards the Employer under the terms of the Contract;

(*b*)   that the Nominated Sub-contractor will save harmless and indemnify the Contractor against all claims demands and proceedings damages costs charges and expenses whatsoever arising out of or in connection with any failure by the Nominated Sub-contractor to perform such obligations or fulfil such liabilities;

(*c*)   that the Nominated Sub-contractor will save harmless and indemnify the Contractor from and against any negligence by the Nominated Sub-contractor his agents workmen and servants and against any misuse by him or them of any Equipment provided by the Contractor for the purposes of the Contract and for all claims as aforesaid;

(*d*)   equivalent to those contained in Clause 63.

Engineer's Action upon Objection

(2)   If pursuant to sub-clause (1) of this Clause the Contractor shall not be obliged to enter into a sub-contract with a Nominated Sub-contractor and shall decline to do so the Engineer shall do one or more of the following:

(*a*)   nominate an alternative sub-contractor in which case sub-clause (1) of this Clause shall apply;

(*b*)   by order under Clause 51 vary the Investigation or the work goods materials or services the subject of the Provisional Sum or Prime Cost Item including if necessary the omission of any such work goods materials or services so that they may be provided by workmen contractors or suppliers as the case may be employed by the Employer either concurrently with the Investigation (in which case Clause 31 shall apply) or at some other date. Provided that in respect of the omission of any Prime Cost Item there shall be included in the Contract Price a sum in respect of the Contractor's charges and profit being a percentage of the estimated value of such work goods materials or services omitted at the rate provided in the Bill of Quantities or inserted in the Appendix to the Form of Tender as the case may be;

(*c*)   subject to the Employer's consent where the Contractor declines to enter into a contract with the Nominated Sub-contractor only on the grounds of unwillingness of the Nominated Sub-contractor to contract on the basis of the provisions contained in paragraphs (a) (b) (c) or (d) of sub-clause (1) of this Clause direct the Contractor to enter into a contract with the Nominated Sub-contractor on such other terms as the Engineer shall specify in which case sub-clause (3) of this Clause shall apply;

(*d*)   in accordance with Clause 58 arrange for the Contractor to execute such work or to supply such goods materials or services.

Direction by Engineer

(3)   If the Engineer shall direct the contractor pursuant to sub-clause (2) of this Clause to enter into a sub-contract which does not contain all the provisions referred to in sub-clause (1) of this Clause:

(*a*)   the Contractor shall not be bound to discharge his obligations and liabilities under the Contract to the extent that the sub-contract terms so specified by the Engineer are inconsistent with the discharge of the same;

(*b*)   in the event of the Contractor incurring loss or expense or suffering damage arising out of the refusal of the Nominated Sub-contractor to accept such provisions the Contractor shall subject to Clause 52(4) be paid in accordance with Clause 60 the amount of such loss or expense or damage as the Contractor could not reasonably avoid.

Contractor Responsible for Nominated Sub-contracts

(4)   Except as otherwise provided in this Clause and in Clause 59B the Contractor shall be as responsible for the work executed or goods materials or services supplied by a Nominated Sub-contractor employed by him as if he had himself executed such work or supplied such goods materials or services or had sub-let the same in accordance with Clause 4.

Payments

(5)   For all work executed or goods materials or services supplied by Nominated Sub-contractors there shall be included in the Contract Price:

(*a*)   the actual price paid or due to be paid by the Contractor in accordance with the terms of the sub-contract (unless and to the extent that any such payment is the result of a default of the Contractor) net of all trade and other discounts rebates and allowances other than any discount obtainable by the Contractor for prompt payment;

146

(*b*)   the sum (if any) provided in the Bill of Quantities for labours in connection therewith or if ordered pursuant to Clause 58(7)(b) as may be determined by the Engineer;

(*c*)   in respect of all other charges and profit a sum being a percentage of the actual price paid or due to be paid calculated (where provision has been made in the Bill of Quantities for a rate to be set against the relevant item of prime cost) at the rate inserted by the Contractor against that item or (where no such provision has been made) at the rate inserted by the Contractor in the Appendix to the Form of Tender as the percentage for adjustment of sums set against Prime Cost Items.

(6)   In the event that the Nominated Sub-contractor shall be in breach of the sub-contract **Breach of Sub-contract** which breach causes the Contractor to be in breach of contract the Employer shall not enforce any award of any arbitrator or judgment which he may obtain against the Contractor in respect of such breach of contract except to the extent that the Contractor may have been able to recover the amount thereof from the Sub-contractor. Provided always that if the Contractor shall not comply with Clause 59B (6) the Employer may enforce any such award or judgment in full.

59B. (1)   Subject to Clause 59A(2)(c) the Contractor shall in every sub-contract with a Nominated **Forfeiture of Sub-** Sub-contractor incorporate provisions equivalent to those provided in Clause 63 and such provi- **contract** sions are hereinafter referred to as 'the Forfeiture Clause'.

(2)   If any event arises which in the opinion of the Contractor would entitle the Contractor to **Termination of Sub-** exercise his right under the Forfeiture Clause (or in the event that there shall be no Forfeiture **contract** Clause in the sub-contract his right to treat the sub-contract as repudiated by the Nominated Sub-contractor) he shall at once notify the Engineer in writing and if he desires to exercise such right by such notice seek the Employer's consent to his so doing. The Engineer shall by notice in writing to the Contractor inform him whether or not the Employer does so consent and if the Engineer does not give notice withholding consent within 7 days of receipt of the Contractor's notice the Employer shall be deemed to have consented to the exercise of the said right. If notice is given by the Contractor to the Engineer under this sub-clause and has not been withdrawn then notwithstanding that the Contractor has not sought the Employer's consent as aforesaid the Engineer may with the Employer's consent direct the Contractor to give notice to the Nominated Sub-contractor expelling the Nominated Sub-contractor from the sub-contract work pursuant to the Forfeiture Clause or rescinding the sub-contract as the case may be. Any such notice given to the Nominated Sub-contractor is hereinafter referred to as a notice enforcing forfeiture of the sub-contract.

(3)   If the Contractor shall give a notice enforcing forfeiture of the sub-contract whether **Engineer's Action upon** under and in accordance with the Forfeiture Clause in the sub-contract or in purported exercise of **Termination** his right to treat the sub-contract as repudiated the Engineer shall do any one or more of the things described in paragraphs (a) (b) and (d) of Clause 59A(2).

(4)   If a notice enforcing forfeiture of the sub-contract shall have been given with the consent **Delay and Extra Cost** of the Employer or by the direction of the Engineer or if it shall have been given without the Employer's consent in circumstances which entitled the Contractor to give such a notice:

(*a*)   there shall be included in the Contract Price:
(i)   the value determined in accordance with Clause 52 of any work the Contractor may have executed or goods or materials he may have provided subsequent to the forfeiture taking effect and pursuant to the Engineer's direction;
(ii)   such amount calculated in accordance with paragraph (a) of Clause 59A (5) as may be due in respect of any work goods materials or services provided by an alternative Nominated Sub-contractor together with reasonable sums for labours and for all other charges and profit as may be determined by the Engineer;
(iii)   any such amount as may be due in respect of the forfeited sub-contract in accordance with Clause 59A (5);
(*b*)   the Engineer shall take any delay to the completion of the Investigation consequent upon the forfeiture into account in determining any extension of time to which the Contractor is entitled under Clause 44 and the Contractor shall subject to Clause 52(4) be paid in accordance with Clause 60 the amount of any additional cost which he may have necessarily and properly incurred as a result of such delay;
(*c*)   the Employer shall subject to Clause 60(7) be entitled to recover from the Contractor upon the certificate of the Engineer issued in accordance with Clause 60(3):
(i)   the amount by which the total sum to be included in the Contract Price pursuant to paragraphs (a) and (b) of this sub-clause exceeds the sum which would but for the forfeiture have been included in the Contract Price in respect of work materials goods and services done supplied or performed under the forfeited sub-contract;
(ii)   all such other loss expense and damage as the Employer may have suffered in consequence of the breach of the sub-contract;

all of which are hereinafter collectively called 'the Employer's loss'.
Provided always that if the Contractor shall show that despite his having complied with sub-clause (6) of this Clause he has been unable to recover the whole or

147

any part of the Employer's loss from the Sub-contractor the Employer shall allow or (if he has already recovered the same from the Contractor) shall repay to the Contractor so much of the Employer's loss as was irrecoverable from the Sub-contractor except and to the extent that the same was irrecoverable by reason of some breach of the sub-contract or other default towards the Sub-contractor by the Contractor or except to the extent that any act or default of the Contractor may have caused or contributed to any of the Employer's loss. Any such repayment by the Employer shall carry interest at the rate stipulated in Clause 60(6) from the date of the recovery by the Employer from the Contractor of the sum repaid.

Termination Without Consent

(5)   If notice enforcing forfeiture of the sub-contract shall have been given without the consent of the Employer and in circumstances which did not entitle the Contractor to give such a notice:

(a)   there shall be included in the Contract Price in respect of the whole of the work covered by the Nominated Sub-contract only the amount that would have been payable to the Nominated Sub-contractor on due completion of the sub-contract had it not been terminated;

(b)   the Contractor shall not be entitled to any extension of time because of such termination nor to any additional expense incurred as a result of the work having been carried out and completed otherwse than by the Sub-contractor;

(c)   the Employer shall be entitled to recover from the Contractor any additional expense he may incur beyond that which he would have incurred had the sub-contract not been terminated.

Recovery of Employer's Loss

(6)   The Contractor shall take all necessary steps and proceedings as may be required by the Employer to enforce the provisions of the sub-contract and/or all other rights and/or remedies available to him so as to recover the Employer's loss from the Sub-contractor. Except in the case where notice enforcing forfeiture of the sub-contract shall have been given without the consent of the Employer and in circumstances which did not entitle the Contractor to give such a notice the Employer shall pay to the Contractor so much of the reasonable costs and expenses of such steps and proceedings as are irrecoverable from the Sub-contractor provided that if the Contractor shall seek to recover by the same steps and proceedings any loss damage or expense additional to the Employer's loss the said irrecoverable costs and expenses shall be borne by the Contractor and the Employer in such proportions as may be fair in all the circumstances.

Payment to Nominated Sub-contractors

59C. Before issuing any certificate under Clause 60 the Engineer shall be entitled to demand from the Contractor reasonable proof that all sums (less retentions provided for in the sub-contract) included in previous certificates in respect of the work executed or goods or materials or services supplied by Nominated Sub-contractors have been paid to the Nominated Sub-contractors or discharged by the Contractor in default whereof unless the Contractor shall:

(a)   give details to the Engineer in writing of any reasonable cause he may have for withholding or refusing to make such payment; and

(b)   produce to the Engineer reasonable proof that he has so informed such Nominated Sub-contractor in writing;

the Employer shall be entitled to pay to such Nominated Sub-contractor direct upon the certification of the Engineer all payments (less retentions provided for in the sub-contract) which the Contractor has failed to make to such Nominated Sub-contractor and to deduct by way of set-off the amount so paid by the Employer from any sums due or which become due from the Employer to the Contractor. Provided always that where the Engineer has certified and the Employer has paid direct as aforesaid the Engineer shall in issuing any further certificate in favour of the Contractor deduct from the amount thereof the amount so paid direct as aforesaid but shall not withhold or delay the issue of the certificate itself when due to be issued under the terms of the Contract.

CERTIFICATES AND PAYMENT

Monthly Statements

60.   (1)   The Contractor shall submit to the Engneer after the end of each month a statement (in such form if any as may be prescribed in the Specification) showing:

(a)   the estimated contract value of the Investigation executed up to the end of that month;

(b)   a list of any goods or materials delivered to the Site for but not yet incorporated in the Ancillary Works and their value;

(c)   the estimated amounts to which the Contractor considers himself entitled in connection with all other matters for which provision is made under the Contract including any Equipment for which separate amounts are included in the Bill of Quantities;

unless in the opinion of the Contractor such values and amounts together will not justify the issue of an interim certificate.

148

Amounts payable in respect of Nominated Sub-contractors are to be listed separately.

(2)   Within 28 days of the date of delivery to the Engineer or Engineer's Representative in accordance with sub-clause (1) of this Clause of the Contractor's monthly statement the Engineer shall certify and the Employer shall pay to the Contractor (after deducting any previous payments on account):

    (*a*)   the amount which in the opinion of the Engineer on the basis of the monthly statement is due to the Contractor on account of sub-clause (1)(a) and (c) of this Clause less a retention as provided in sub-clause (4) of this Clause;

    (*b*)   such amounts (if any) as the Engineer may consider proper (but in no case exceeding the percentage of the value stated in the Appendix to the Form of Tender) in respect of (b) of sub-clause (1) of this Clause which amounts shall not be subject to a retention under sub-clause (4) of this Clause.

The amounts certified in respect of Nominated Sub-contracts shall be shown separately in the certificate. The Engineer shall not be bound to issue an interim certificate for a sum less than that named in the Appendix to the Form of Tender but shall not on that account be entitled to defer the issue of an interim certificate beyond the date for certification of the Contractor's next monthly statement.

(3)   Not later than 3 months after the date of the Acceptance Certificate the Contractor shall submit to the Engineer a statement of final account and supporting documentation showing in detail the value in accordance with the Contract of the work done in accordance with the Contract together with all further sums which the Contractor considers to be due to him under the Contract up to the date of the Acceptance Certificate. Within 3 months after receipt of this final account and of all information reasonably required for its verification the Engineer shall issue a final certificate stating the amount which in his opinion is finally due under the Contract up to the date of the Acceptance Certificate and after giving credit to the Employer for all amounts previously paid by the Employer and for all sums to which the Employer is entitled under the Contract up to the date of the Acceptance Certificate the balance if any due from the Employer to the Contractor or from the Contractor to the Employer as the case may be. Such balance shall subject to Clause 47 be paid to or by the Contractor as the case may require within 28 days of the date of the certificate.

(4)   If the Tender shall contain a requirement (stipulated in the Appendix to the Form of Tender) for a retention to be made pursuant to sub-clause (2)(a) of this Clause it shall be a sum equal to 5 per cent of the amount due to the Contractor until a reserve shall have accumulated in the hand of the Employer up to the following limits:

    (*a*)   where the Tender Total does not exceed £50,000 5 per cent of the Tender Total but not exceeding £1,500; or

    (*b*)   where the Tender Total exceeds £50,000 3 per cent of the Tender Total;

except that the limit shall be reduced by the amount of any payment that shall have been made pursuant to sub-clause (5) of this Clause.

(5)   (*a*)   If the Engineer shall issue a Certificate of Completion in respect of any Section or part of the Site Operations pursuant to Clause 48(2) or (3) there shall become due on the date of issue of such certificate and shall be paid to the Contractor within 14 days thereof a sum equal to 1½ per cent of the amount due to the Contractor at that date in respect of such Section or part as certified for payment pursuant to sub-clause (2) of this Clause provided that any sum or sums paid under this sub-clause shall not exceed in aggregate one half of the retention money deducted in accordance with sub-clause 2(a) of this Clause from the payments made to the Contractor at the date of issue by the Engineer of the aforesaid Certificate of Completion.

    (*b*)   One half of the retention money less any sums paid pursuant to sub-clause (5)(a) of this Clause shall be paid to the Contractor within 14 days after the date on which the Engineer shall have issued a Certificate of Completion for the whole of the Site Operations pursuant to Clause 48(2).

    (*c*)   The other half of the retention money shall be paid to the Contractor within 14 days after the expiration of the Period of Maintenance notwithstanding that at such time there may be outstanding claims by the Contractor against the Employer. Provided always that if at such time there shall remain to be executed by the Contractor any outstanding work referred to under Clause 48 or any work ordered during such period pursuant to Clauses 49 and 50 the Employer shall be entitled to withhold payment until the completion of such work of so much of the second half of retention money as shall in the opinion of the Engineer represent the cost of the work so remaining to be executed.

       Provided further that in the event of different maintenance periods having become applicable to different Sections or parts of the Site Operations pursuant to Clause 48 the expression 'expiration of the Period of Maintenance' shall for the purposes of this sub-clause be deemed to mean the expiration of the latest of such periods.

Monthly Payments

Final Account

Retention

Payment of Retention Money

149

Interest on Overdue Payments

(6)   In the event of failure by the Engineer to certify or the Employer to make payment in accordance with sub-clauses (2) (3) and (5) of this Clause the Employer shall pay to the Contractor interest upon any payment overdue thereunder at a rate per annum equivalent to 2 per cent above the Average Base Lending Rate of Lloyds Barclays National Westminster and Midland Banks current on the date upon which payment first becomes overdue. In the event of any variation in the said Average Base Lending Rate being announced whilst such payment remains overdue the interest payable to the Contractor for the period that such payment remains overdue shall be correspondingly varied from the date of each such variation.

Correction and Withholding of Certificates

(7)   The Engineer shall have power to omit from any certificate the value of any work done goods or materials supplied or services rendered with which he may for the time being be dissatisfied and for that purpose or for any other reason which to him may seem proper may by any certificate delete correct or modify any sum previously certified by him. Provided always that:

(*a*)   the Engineer shall not in any interim certificate delete or reduce any sum previously certified in respect of work done goods or materials supplied or services rendered by a Nominated Sub-contractor if the Contractor shall have already paid or be bound to pay that sum to the Nominated Sub-contractor;

(*b*)   if the Engineer in the final certificate shall delete or reduce any sum previously certified in respect of work done goods or materials supplied or services rendered by a Nominated Sub-contractor which sum shall have been already paid by the Contractor to the Nominated Sub-contractor the Employer shall reimburse to the Contractor the amount of any sum overpaid by the Contractor to the Sub-contractor in accordance with the certificates issued under sub-clause (2) of this Clause which the Contractor despite compliance with Clause 59B(6) shall be unable to recover from the Nominated Sub-contractor together with interest thereon at the rate stated in Clause 60(6) from 28 days after the date of the final certificate issued under sub-clause (3) of this Clause until the date of such reimbursement.

Copy of Certificate for Contractor

(8)   Every certificate issued by the Employer pursuant to this Clause shall be sent to the Employer and at the same time a copy thereof shall be sent to the Contractor.

Acceptance Certificate

61.   (1)   Upon the expiration of the Period of Maintenance or where there is more than one such period upon the expiration of the latest period and when all outstanding work referred to under Clause 48 and all work of repair amendment reconstruction rectification and making good of defects imperfections shrinkages damage subsidence of backfill and other faults referred to under Clauses 49 and 50 shall have been completed and when Completion Certificates shall have been issued in respect of all Laboratory Testing and all Reports (if any) included in the Investigation the Engineer shall issue to the Employer (with a copy to the Contractor) an Acceptance Certificate stating the date on which the Contractor shall have completed his obligations to carry out the Investigation to the Engineer's satisfaction.

Unfulfilled Obligations

(2)   The issue of the Acceptance Certificate shall not be taken as relieving either the Contractor or the Employer from any liability the one towards the other arising out of or in any way connected with the performance of their respective obligations under the Contract.

REMEDIES AND POWERS

Urgent Repairs

62.   If by reason of any accident or failure or other event occurring to in or in connection with the Site Operations or any part thereof either during the execution of the Site Operations or during the Period of Maintenance any remedial or other work or repair shall in the opinion of the Engineer be urgently necessary and the Contractor is unable or unwilling at once to do such work or repair the Employer may by his own or other workmen do such work or repair as the Engineer may consider necessary. If the work or repair so done by the Employer is work which in the opinion of the Engineer the Contractor was liable to do at his own expense under the Contract all costs and charges properly incurred by the Employer in so doing shall on demand be paid by the Contractor to the Employer or may be deducted by the Employer from any monies due or which may become due to the Contractor. Provided always that the Engineer shall as soon after the occurrence of any such emergency as may be reasonably practicable notify the Contractor thereof in writing.

Forfeiture

63.   (1)   If the Contractor shall become bankrupt or have a receiving order made against him or shall present his petition in bankruptcy or shall make an arrangement with or assignment in favour of his creditors or shall agree to carry out the Contract under a committee of inspection of his creditors or (being a corporation) shall go into liquidation (other than a voluntary liquidation for the purposes or amalgamation or reconstruction) or if the Contractor shall assign the Contract without the consent in writing of the Employer first obtained or shall have an execution levied on his goods or if the Engineer shall certify in writing to the Employer that in his opinion the Contractor:

(*a*)   has abandoned the Contract; or

(*b*)   without reasonable excuse has failed to commence the Investigation in accordance with Clause 41 or has suspended the progress of the Investigation for 14 days after receiving from the Engineer written notice to proceed; or

150

(c) has failed to remove goods or materials from the Site or to rectify or to pull down and replace work for 14 days after receiving from the Engineer written notice that the said goods materials or work have been condemned and rejected by the Engineer; or

(d) despite previous warning by the Engineer in writing is failing to proceed with the Investigation with due diligence or is otherwise persistently or fundamentally in breach of his obligations under the Contract; or

(e) has to the detriment of good workmanship or in defiance of the Engineer's instruction to the contrary sub-let any part of the Contract;

then the Employer may after giving 7 days' notice in writing to the Contractor enter upon the Site and expel the Contractor therefrom without thereby avoiding the Contract or releasing the Contractor from any of his obligations or liabilities under the Contract or affecting the rights and powers conferred on the Employer or the Engineer by the Contract and may himself complete the Investigation or may employ any other contractor to complete the Investigation and the Employer or such other contractor may use for such completion so much of the Equipment goods and materials which have been deemed to become the property of the Employer under Clause 53 as he or they may think proper and the Employer may at any time sell any of the said Equipment and unused goods and materials and apply the proceeds of sale in or towards the satisfaction of any sums due or which may become due to him from the Contractor under the Contract.

(2) By the said notice or by further notice in writing within 14 days of the date thereof the Engineer may require the Contractor to assign to the Employer and if so required the Contractor shall forthwith assign to the Employer the benefit of any agreement for the supply of any goods or materials and/or for the execution of any work for the purposes of this Contract which the Contractor may have entered into.

Assignment to Employer

(3) The Engineer shall as soon as may be practicable after any such entry and expulsion by the Employer fix and determine *ex parte* or by or after reference to the parties or after such investigation or enquiries as he may think fit to make or institute and shall certify what amount (if any) had at the time of such entry and expulsion been reasonably earned by or would reasonably accrue to the Contractor in respect of work then actually done by him under the Contract and what was the value of any unused or partially used goods and materials and Equipment which have been deemed to become the property of the Employer under Clause 53.

Valuation at Date of Forfeiture

(4) If the Employer shall enter and expel the Contractor under this Clause he shall not be liable to pay to the Contractor any money on account of the Contract until the expiration of the Period of Maintenance and thereafter until the costs of completion and maintenance damages for delay in completion (if any) and all other expenses incurred by the Employer have been ascertained and the amount thereof certified by the Engineer. The Contractor shall then be entitled to receive only such sum or sums (if any) as the Engineer may certify would have been due to him upon due completion by him after deducting the said amount. But if such amount shall exceed the sum which would have been payable to the Contractor on due completion by him then the Contractor shall upon demand pay to the Employer the amount of such excess and it shall be deemed a debt due by the Contractor to the Employer and shall be recoverable accordingly.

Payment after Forfeiture

FRUSTRATION

64. In the event of the Contract being frustrated whether by war or by any other supervening event which may occur independently of the will of the parties the sum payable by the Employer to the Contractor in respect of the work executed shall be the same as that which would have been payable under Clause 65(5) if the Contract had been determined by the Employer under Clause 65.

Payment in Event of Frustration

WAR CLAUSE

65. (1) If during the currency of the Contract there shall be an outbreak of war (whether war is declared or not) in which Great Britain shall be engaged on a scale involving general mobilisation of the armed forces of the Crown the Contractor shall for a period of 28 days reckoned from midnight on the date that the order for general mobilisation is given continue so far as is physically possible to execute the Investigation in accordance with the Contract.

Works to Continue for 28 Days on Outbreak of War

(2) If at any time before the expiration of the said period of 28 days the Investigation shall have been completed or completed so far as to be usable all provisions of the Contract shall continue to have full force and effect save that:

Effect of Completion Within 28 Days

(a) the Contractor shall in lieu of fulfilling his obligations under Clauses 49 and 50 be entitled at his option to allow against the sum due to him under the provisions hereof the cost (calculated at the prices ruling at the beginning of the said period of 28 days) as certified by the Engineer at the expiration of the Period of Maintenance of repair rectification and making good any work for the repair rectification or making good of which the Contractor would have been liable under the said Clauses had they continued to be applicable;

151

(*b*) the Employer shall not be entitled at the expiration of the Period of Maintenance to withhold payment under Clause 60(5)(c) of the second half of the retention money or any part thereof except such sum as may be allowable under the Contractor under the provisions of the last preceding paragraph which sum may (without prejudice to any other mode of recovery thereof) be deducted by the Employer from such second half.

Right of Employer to Determine Contract

(3)   If the Investigation shall not have been completed as aforesaid the Employer shall be entitled to determine the Contract (with the exception of this Clause and Clauses 66 and 68) by giving notice in writing to the Contractor at any time after the aforesaid period of 28 days has expired and upon such notice being given the Contract shall (except as above mentioned) forthwith determine but without prejudice to the claims of either party in respect of any antecedent breach thereof.

Removal of Equipment on Determination

(4)   If the Contract shall be determined under the provisions of the last preceding sub-clause the Contractor shall with all reasonable despatch remove from the Site all his Equipment and shall give facilities to his sub-contractors to remove similarly all Equipment belonging to them and in the event of any failure so to do the Employer shall have the like powers as are contained in Clause 53(8) in regard to failure to remove Equipment on completion of the Works but subject to the same condition as is contained in Clause 53(9).

Payment on Determination

(5)   If the Contract shall be determined as aforesaid the Contractor shall be paid by the Employer (insofar as such amounts or items shall not have been already covered by payment on account made to the Contractor) for all work executed prior to the date of determination at the rates and prices provided in the Contract and in addition:

(*a*) the amounts payable in respect of any preliminary items so far as the work or service comprised therein has been carried out or performed and a proper proportion as certified by the Engineer of any such items the work or service comprised in which has been partially carried out or performed;

(*b*) the cost of materials or goods reasonably ordered for the Investigation which shall have been delivered to the Contractor or of which the Contractor is legally liable to accept delivery (such materials or goods becoming the property of the Employer upon such payment being made by him);

(*c*) a sum to be certified by the Engineer being the amount of any expenditure reasonably incurred by the Contractor in the expectation of completing the whole of the Investigation insofar as such expenditure shall not have been covered by the payments in this sub-clause before mentioned;

(*d*) any additional sum payable under sub-clause (6)(b)(c) and (d) of this Clause;

(*e*) the reasonable cost of removal under sub-clause (4) of this Clause.

Provisions to Apply as from Outbreak of War

(6)   Whether the Contract shall be determined under the provisions of sub-clause (3) of this Clause or not the following provisions shall apply or be deemed to have applied as from the date of the said outbreak of war notwithstanding anything expressed in or implied by the other terms of the Contract viz.:

(*a*) The Contractor shall be under no liability whatsoever whether by way of indemnity or otherwise for or in respect of damage to the Site Operations or damage or loss of any samples or cores obtained or records or reports made in the course thereof or to property (other than property of the Contractor or property hired by him for the purposes of executing the Investigation) whether of the Employer or of third parties or for or in respect of injury or loss of life to persons which is the consequence whether direct or indirect of war hostilities (whether war has been declared or not) invasion act of the Queen's enemies civil war rebellion revolution insurrection military or usurped power and the Employer shall indemnify the Contractor against all such liabilities and against all claims demands proceedings damages costs charges and expenses whatsoever arising thereout or in connection therewith.

(*b*) If the Site Operations or the said samples cores records or reports shall sustain destruction or are lost or damaged by reason of any of the causes mentioned in the last preceding paragraph the Contractor shall nevertheless be entitled to payment for any part of the Site Operations samples cores records or reports so destroyed lost or damaged and the Contractor shall be entitled to be paid by the Employer the cost of making good any such destruction or damage so far as may be required by the Engineer or as may be necessary for the completion of the Investigation on a cost basis plus such profit as the Engineer may certify to be reasonable.

(*c*) In the event that the Contract includes the Contract Price Fluctuations Clause the terms of that Clause shall continue to apply but if subsequent to the outbreak of war the index figures therein referred to shall cease to be published or in the event that the contract shall not include a Price Fluctuations Clause in that form the following paragraph shall have effect:

   If under the decision of the Civil Engineering Construction Conciliation Board or of any other body recognised as an appropriate body for regulating the rates of wages in any trade or industry other than the Civil Engineering Construction Industry to which Contractors undertaking works of civil engineering construction give effect by agreement or in practice or by reason of any Statute or Statutory Instrument there shall during the currency of the Contract be any increase or decrease in the wages or the rates of wages or in the allowances or rates of allowances (including allowances in respect of holidays) payable to or in respect of labour of any kind prevailing at the date of outbreak of war as then fixed by the said Board or such other body as aforesaid or by Statute or Statutory Instrument or any increase in the amount payable by the Contractor by virtue or in respect of any Scheme of State Insurance or if there shall be any increase or decrease in the cost prevailing at the date of the said outbreak of war of any materials consumable stores fuel or power which increase or increases decrease or decreases shall result in an increase or decrease of cost to the Contractor in carrying out the Investigation the net increase or decrease of cost shall form an addition or deduction as the case may be to or from the Contract Price and be paid to or allowed by the Contractor accordingly.

(*d*) If the cost of the Investigation to the Contractor shall be increased or decreased by reason of the provisions of any Statute or Statutory Instrument or other Government or Local Government Order or Regulation becoming applicable to the Investigation after the date of the said outbreak of war or by reason of any trade or industrial agreement entered into after such date to which the Civil Engineering Construction Conciliation Board or any other body as aforesaid is party or gives effect or by reason of any amendment of whatsoever nature of the Working Rule Agreement of the said Board or of any other body as aforesaid or by reason of any other circumstance or thing attributable to or consequent on such outbreak of war such increase or decrease of cost as certified by the Engineer shall be reimbursed by the Employer to the Contractor or allowed by the Contractor as the case may be.

(*e*) Damage or injury caused by the explosion whenever occurring of any mine bomb shell grenade or other projectile missile or munition of war and whether occurring before or after the cessation of hostilities shall be deemed to be the consequence of any of the events mentioned in sub-clause (6)(a) of this Clause.

SETTLEMENT OF DISPUTES

66. (1) If any dispute or difference of any kind whatsoever shall arise between the Employer and the Contractor in connection with or arising out of the Contract or the carrying out of the Investigation including any dispute as to any decision opinion instruction direction certificate or valuation of the Engineer (whether during the progress of the Investigation or after its completion and whether before or after the determination abandonment or breach of the Contract) it shall be referred to and settled by the Engineer who shall state his decision in writing and give notice of the same to the Employer and the Contractor. Unless the Contract shall have been already determined or abandoned the Contractor shall in every case continue to proceed with the Investigation with all due diligence and he shall give effect forthwith to every such decision of the Engineer unless and until the same shall be revised by an arbitrator as hereinafter provided. Such decisions shall be final and binding upon the Contractor and the Employer unless either of them shall require that the matter be referred to arbitration as hereinafter provided. If the Engineer shall fail to give such decision for a period of 3 calendar months after being requested to do so or if either the Employer or the Contractor be dissatisfied with any such decision of the Engineer then and in any such case either the Employer or the Contractor may within 3 calendar months after receiving notice of such decision or within 3 calendar months after the expiration of the said period of 3 months (as the case may be) require that the matter shall be referred to the arbitration of a person to be agreed upon between the parties or (if the parties fail to appoint an arbitrator within one calendar month of either party serving on the other party a written notice to concur in the appointment of an arbitrator) a person to be appointed on the application of either party by the President for the time being of the Institution of Civil Engineers. If an arbitrator declines the appointment or after appointment is removed by order of a competent court or is incapable of acting or dies and the parties do not within one calendar month of the vacancy arising fill the vacancy then the President for the time being of the Institution of Civil Engineers may on the application of either party appoint an arbitrator to fill the vacancy. Any such reference to arbitration shall be deemed to be a submission to arbitration within the meaning of the Arbitration Acts 1950–1979 or the Arbitration (Scotland) Act 1894 as the case may be or any statutory re-enactment or amendment thereof for the time being in force. Any such reference to arbitration may be conducted in accordance with the Institution of Civil Engineers' Arbitration Procedure (1983) or any amendment or modification thereof being in force at the time of the appointment of the arbitrator and in cases where the President of the Institution of Civil Engineers is requested to appoint the arbitrator he may direct that the arbitration is conducted in accordance with the aforementioned Procedure or any amendment or modification thereof. Such arbitrator shall have full power to open up review and revise any decision opinion instruction direction certificate or valuation of the Engineer and neither party shall be limited in the proceedings before such arbitrator

Settlement of
Disputes—Arbitration

to the evidence or arguments put before the Engineer for the purpose of obtaining his decision above referred to. The award of the arbitrator shall be final and binding on the parties. Save as provided for in sub-clause (2) of this Clause no steps shall be taken in the reference to the arbitrator until after the completion or alleged completion of the Investigation unless with the written consent of the Employer and the Contractor. Provided always:

(*a*) that the giving of a Certificate of Completion under Clause 48 shall not be a condition precedent to the taking of any step in such reference;

(*b*) that no decision given by the Engineer in accordance with the foregoing provisions shall disqualify him from being called as a witness and giving evidence before the arbitrator on any matter whatsoever relevant to the dispute or difference so referred to the arbitrator as aforesaid.

Interim Arbitration

(2) In the case of any dispute or difference as to any matter arising under Clause 12 or the withholding by the Engineer of any certificate or the withholding of any portion of the retention money under Clause 60 to which the Contractor claims to be entitled or as to the exercise of the Engineer's power to give a certificate under Clause 63(1) the reference to the arbitrator may proceed notwithstanding that the Investigation shall not then be or be alleged to be complete.

Vice-President to Act

(3) In any case where the President for the time being of the Institution of Civil Engineers is not able to exercise the functions conferred on him by this Clause the said functions may be exercised on his behalf by a Vice-President for the time being of the said Institution.

APPLICATION TO SCOTLAND

Application to Scotland

67. If the Site is situated in Scotland the Contract shall in all respects be construed and operate as a Scottish contract and shall be interpreted in accordance with Scots law.

NOTICES

Service of Notice on Contractor

68. (1) Any notice to be given to the Contractor under the terms of the Contract shall be served by sending the same by post to or leaving the same at the Contractor's principal place of business (or in the event of the Contractor being a Company to or at its registered office).

Service of Notice on Employer

(2) Any notice to be given to the Employer under the terms of the Contract shall be served by sending the same by post to or leaving the same at the Employer's last known address (or in the event of the Employer being a Company to or at its registered office).

TAX MATTERS

Tax Fluctuations

69. (1) The rates and prices contained in the Bill of Quantities take account of the levels and incidence at the date for return of tenders (hereinafter called 'the relevant date') of the taxes levies and contributions (including national insurance contributions but excluding income tax and any levy payable under the Industrial Training Act 1964) which are by law payable by the Contractor in respect of his workpeople and the premiums and refunds (if any) which are by law payable to the Contractor in respect of his workpeople. Any such matter is hereinafter called 'a labour-tax matter'.

The rates and prices contained in the Bill of Quantities do not take account of any level or incidence of the aforesaid matters where at the relevant date such level or incidence does not then have effect but although then known is to take effect at some later date. The taking effect of any such level or incidence at the later date shall for the purposes of sub-clause (2) of this Clause be treated as the occurrence of an event.

(2) If after the relevant date there shall occur any of the events specified in sub-clause (3) of this Clause and as a consequence thereof the cost to the Contractor of performing his obligations under the Contract shall be increased or decreased then subject to the provisions of sub-clause (4) of this Clause the net amount of such increase or decrease shall constitute an addition to or deduction from the sums otherwise payable to the Contractor under the Contract as the case may require.

(3) The events referred to in the preceding sub-clause are as follows:

(*a*) any change in the level of any labour-tax matter;

(*b*) any change in the incidence of any labour-tax matter including the imposition of any new such matter or the abolition of any previously existing such matter.

(4) In this Clause workpeople means persons employed by the Contractor on manual labour whether skilled or unskilled but for the purpose of ascertaining what if any additions or deductions are to be paid or allowed under this Clause account shall not be taken of any labour-tax matter in relation to any workpeople of the Contractor unless at the relevant time their normal place of employment is the Site.

(5) Subject to the provisions of the Contract as to the placing of sub-contracts with Nominated Sub-contractors the Contractor may incorporate in any sub-contract made for the purpose of performing his obligations under the Contract provisions which are *mutatis mutandis* the same as the provisions of this Clause and in such event additions or deductions to be made in ac-

cordance with any such sub-contract shall also be made under the Contract as if the increase or decrease of cost to the sub-contractor had been directly incurred by the Contractor.

(6)   As soon as practicable after the occurrence of any of the events specified in sub-clause (3) of this Clause the Contractor shall give the Engineer notice thereof. The Contractor shall keep such contemporary records as are necessary for the purpose of ascertaining the amount of any addition or deduction to be made in accordance with this Clause and shall permit the Engineer to inspect such records. The Contractor shall submit to the Engineer with his monthly statements full details of every addition or deduction to be made in accordance with this Clause. All certificates for payment issued after submission of such details shall take due account of the additions or deductions to which such details relate. Provided that the Engineer may if the Contractor fails to submit full details of any deduction nevertheless take account of such deduction when issuing any certificate for payment.

70.   (1)   In this Clause 'exempt supply' 'invoice' 'tax' 'taxable person' and 'taxable supply' have **Value Added Tax** the same meanings as in Part I of the Finance Act 1972 (hereinafter referred to as 'the Act') including any amendment or re-enactment thereof and any reference to the Value Added Tax (General) Regulations 1972 (SI 1972/1147) (hereinafter referred to as the VAT Regulations) shall be treated as a reference to any enactment corresponding to those regulations for the time being in force in consequence of any amendment or re-enactment of those regulations.

(2)   The Contractor shall be deemed not to have allowed in his tender for the tax payable by him as a taxable person to the Commissioners of Customs and Excise being tax chargeable on any taxable supplies to the Employer which are to be made under the Contract.

(3)   (a)   The Contractor shall not in any statement submitted under Clause 60 include any element on account of tax in any item or claim contained in or submitted with the statement.

(b)   The Contractor shall concurrently with the submission of the statement referred to in sub-clause (3)(a) of this Clause furnish the Employer with a written estimate showing those supplies of goods and services and the values thereof included in the said statement and on which tax will be chargeable under Regulation 21 of the VAT Regulations at a rate other than zero.

(4)   At the same time as payment (other than payment in accordance with this sub-clause) for goods or services which were the subject of a taxable supply provided by the Contractor as a taxable person to the Employer is made in accordance with the Contract there shall also be paid by the Employer a sum (separately identified by the Employer and in this Clause referred to as 'the tax payment') equal to the amount of tax payable by the Contractor on that supply. Within seven days of each payment the Contractor shall:

(a)   if he agrees with that tax payment or any part thereof issue to the Employer an authenticated receipt of the kind referred to in Regulation 21(2) of the VAT Regulations in respect of that payment or that part; and

(b)   if he disagrees with that tax payment or any part thereof notify the Employer in writing stating the grounds of his disagreement.

(5)   (a)   If any dispute difference or question arises between the Employer and the Contractor in relation to any of the matters specified in Section 40(1) of the Act then:

(i)   if the Employer so requires the Contractor shall refer the matter to the said Commissioners for their decision on it

(ii)   if the Contractor refers the matter to the said Commissioners (whether or not in pursuance of sub-paragraph (i) above) and the Employer is dissatisfied with their decision on the matter the Contractor shall at the Employer's request refer the matter to a Value Added Tax Tribunal by way of appeal under Section 40 of the Act whether the Contractor is so dissatisfied or not.

(iii)   a sum of money equal to the amount of tax which the Contractor in making a deposit with the said Commissioners under Section 40(3)(a) of the Act is required so to deposit shall be paid to the Contractor; and

(iv)   if the Employer requires the Contractor to refer such a matter to the Tribunal in accordance with sub-paragraph (ii) above then he shall reimburse the Contractor any costs and any expenses reasonably and properly incurred in making that reference less any costs awarded to the Contractor by the Tribunal and the decision of the Tribunal shall be binding on the Employer to the same extent as it binds the Contractor.

(b)   Clause 66 shall not apply to any dispute difference or question arising under paragraph (a) of this sub-clause.

(6)   (a)   The Employer shall without prejudice to his rights under any other Clause hereof be entitled to recover from the Contractor:

(i)   any tax payment made to the Contractor of a sum which is in excess of the sum (if any) which in all the circumstances was due in accordance with sub-clause (4) of this Clause

(ii) in respect of any sum of money deposited by the Contractor pursuant to sub-clause (5)(a)(iii) of this Clause a sum equal to the amount repaid under Section 40(4) of the Act together with any interest thereon which may have been determined thereunder.

(b) If the Contractor shall establish that the Commissioners have charged him in respect of a taxable supply for which he has received payment under this Clause tax greater in amount than the sum paid to him by the Employer the Employer shall subject to the provisions of sub-clause (5) of this Clause pay to the Contractor a sum equal to the difference between the tax previously paid and the tax charged to the Contractor by the Commissioners.

(7) If after the date for return of tenders the descriptions of any supplies of goods or services which at the date of tender are taxable or exempt supplies are with effect after the date for return of tenders modified or extended by or under the Act and that modification or extension shall result in the Contractor having to pay either more or less tax or greater or smaller amounts attributable to tax and that tax or those amounts as the case may be shall be a direct expense or direct saving to the Contractor in carrying out the Investigation and not recoverable or allowable under the Contract or otherwise then there shall be paid to or allowed by the Contractor as appropriate a sum equivalent to that tax or amounts as the case may be.

Provided always that before that tax is included in any payment by the Employer or those amounts are included in any certificate by the Engineer as the case may be the Contractor shall supply all the information the Engineer requires to satisfy himself as to the Contractor's entitlement under this sub-clause.

(8) The Contractor shall upon demand pay to the Employer the amount of any sum in accordance with sub-clauses (6) and (7) of this Clause and it shall be deemed a debt due by the Contractor to the Employer and shall be recoverable accordingly.

METRICATION

Metrication

71. (1) If any materials described in the Contract or ordered by the Engineer are described by dimensions in the metric or imperial measure and having used his best endeavours the Contractor cannot without undue delay or additional expense or at all procure such materials in the measure specified in the Contract but can obtain such materials in the other measure to dimensions approximating to those described in the Contract or ordered by the Engineer then the Contractor shall forthwith give written notice to the Engineer of these facts stating the dimensions to which such materials are procurable in the other measure. Such notice shall where practicable be given in sufficient time to enable the Engineer to consider and if necessary give effect to any design change which may be required and to avoid delay in the performance of the Contractor's other obligations under the Contract. Any additional cost or expense incurred by the Contractor as a result of any delay arising out of the Contractor's default under this sub-clause shall be borne by the Contractor.

(2) As soon as practicable after the receipt of any such notice under the preceding sub-clause the Engineer shall if he is satisfied that the Contractor has used his best endeavours to obtain materials to the dimensions described in the Contract or ordered by the Engineer and that they are not obtainable without undue delay or without putting the Contractor to additional expense either:

(a) instruct the Contractor pursuant to Clause 13 to supply such materials (despite such delay or expense) in the dimensions described in the Contract or originally ordered by the Engineer; or

(b) give an order to the Contractor pursuant to Clause 51:
(i) to supply such materials to the dimensions stated in his said notice to be procurable instead of to the dimensions described in the Contract or originally ordered by the Engineer; or
(ii) to make some other variation whereby the need to supply such materials to the dimensions described in the Contract or originally ordered by the Engineer will be avoided.

(3) This Clause shall apply irrespective of whether the materials in question are to be supplied in accordance with the Contract directly by the Contractor or indirectly by a Nominated Subcontractor.

SPECIAL CONDITIONS

Special Conditions

72. The following special conditions form part of the Conditions of Contract.

(Note: Any special conditions which it is desired to incorporate in the conditions of contract should be numbered consecutively with the foregoing conditions of contract).

156

SHORT DESCRIPTION OF INVESTIGATION:

All Site Operations Laboratory Testing* and Report preparation and submission* in connection

with† . . . .

. . . .

FORM OF TENDER

(NOTE: The Appendix forms part of the Tender)

to . . . .

. . . .

. . . .

GENTLEMEN,

Having examined the Drawings, Conditions of Contract, Specification, Schedules and Bill of Quantities for the above-mentioned Investigation (and the matters set out in the Appendix hereto), we offer to carry out the whole of the said Investigation in conformity with the said Drawings, Conditions of Contract, Specification, Schedules and Bill of Quantities for such sum as may be ascertained in accordance with the said Conditions of Contract.

We undertake to complete the whole Investigation comprised in the Contract within the time stated in the Appendix hereto.

If our tender is accepted we will, when required, provide two good and sufficient sureties or obtain the guarantee of a Bank or Insurance Company (to be approved in either case by you) to be jointly and severally bound with us in a sum equal to the percentage of the Tender Total as defined in the said Conditions of Contract for the due performance of the Contract under the terms of a Bond in the form annexed to the Conditions of Contract.

Unless and until a formal Agreement is prepared and executed this Tender, together with your written acceptance thereof, shall constitute a binding Contract between us.

We understand that you are not bound to accept the lowest or any tender you may receive.
* To the best of our knowledge and belief we have complied with the general conditions required by the Fair Wages Resolution for the three months immediately preceding the date of this Tender.

We are, Gentlemen,

Yours faithfully,

Signature . . . .

Address . . . .

. . . .

Date . . . .

* Delete if not required
† Complete as appropriate

157

FORM OF TENDER (APPENDIX)

Appendix

NOTE: Relevant Clause numbers are shown in brackets following the description

Amount of Bond (if any) (10) . . . . % of Tender Total
Period for Approvals (14(5))
Testing Schedule . . . . weeks
Draft Report . . . . weeks
Final Report . . . . weeks
Minimum Amount of Insurance (23(2)) £ . . . .
Time for Completion (43) Liquidated Damages for Delay (47)
Column 1
(see Clause 47(1))

For the Whole of the Investigation

_____ (a) weeks £ . . . . (b) per Day/Week (c)
Column 2 Column 3
(see Clause 47(2))

For the following Sections (of the Investigation) £ . . . . £ . . . .
Section (d) . . . .

. . . . _____Weeks per Day/Week (c) per Day/Week (c)

Section (d) . . . . £ . . . . £ . . . .

. . . . _____Weeks Per Day/Week (c) per Day/Week (c)

Section (d) . . . . £ . . . . £ . . . .
. . . . _____Weeks Per Day/Week (c) per Day/Week (c)

For the whole of the Site Operations

_____ (a) weeks £ . . . . (b) per Day/Week (c)

For the following Sections (of the Site Operations)
Section (d) . . . . £ . . . . £ . . . .

. . . . _____Weeks per Day/Week (c) per Day/Week (c)

Section (d) . . . . £ . . . . £ . . . .

. . . . _____Weeks per Day/Week (c) per Day/Week (c)

Section (d) . . . . £ . . . . £ . . . .

. . . . _____Weeks per Day/Week (c) per Day/Week (c)

Period of Maintenance (49(1)) . . . . Weeks
Standard Method of Measurement adopted in preparation of Bills of Quantities (57)(e) . . . .
. . . .

Percentage for adjustment of P.C. Sums (59A (2)(b) and (5)(c)) . . . . %
Percentage of the Value of Goods and
Materials to be included in Interim
Certificates (60(2)(b) . . . . %
Minimum Amount of Interim Certificates (60(2)) £ . . . .

Retention in accordance with Clause 60(4) is/ is not(g) to be deducted.

(a) To be completed in every case by Contractor if not already stipulated).
(b) To be completed by Engineer in every case.
(c) Delete which not required.
(d) To be completed by the Engineer as required, with brief description.
    The division of the Investigation into Sections (see Clause 1(1)(o)) may be
    based on sub-division of the Site Operations or on separation of the Site
    Operations from the Laboratory Testing and Report or otherwise.
(e) Insert here any amendment or modification adopted if different from that
    stated in Clause 57.
(g) Delete as appropriate.

FORM OF AGREEMENT

THIS AGREEMENT made the . . . . day of . . . .

19 . . . . . . BETWEEN . . . .

of . . . . in the

County of . . . . (hereinafter called 'the Employer') of the one part and

. . . . of . . . .

in the County of . . . .

. . . . (hereinafter called 'the Contractor') of the other part

WHEREAS the Employer is desirous that a certain Investigation should be carried out, viz. the Site

Operations Laboratory Testing* and Report preparation and submission* in connection with . . . . .

. . . . and has accepted a Tender by the

Contractor for the Investigation.

NOW THIS AGREEMENT WITNESSETH as follows:

1. In this Agreement words and expressions shall have the same meanings as are respectively assigned to them in the Conditions of Contract hereinafter referred to.

2. The following documents shall be deemed to form and be read and construed as part of this Agreement, viz.:

    (a)   The said Tender.
    (b)   The Drawings.
    (c)   The Conditions of Contract.
    (d)   The Specification.
    (e)   The Schedules.
    (f)   The Priced Bill of Quantities.

3. In consideration of the payments to be made by the Employer to the Contractor as hereinafter mentioned the Contractor hereby covenants with the Employer to carry out the Investigation in conformity in all respects with the provisions of the Contract.

4. The Employer hereby covenants to pay to the Contractor in consideration of the carrying out completion and maintenance of the Investigation the Contract Price at the times and in the manner prescribed by the Contract.

IN WITNESS whereof the parties hereto have caused their respective Common Seals to be hereunto affixed (or have hereunto set their respective hands and seals) the day and year first above written.

The Common Seal of . . . .

. . . . Limited

was hereunto affixed in the presence of:

or

SIGNED SEALED AND DELIVERED by the

said . . . .

. . . .

in the presence of:

. . . .

. . . .

* Delete if not required

FORM OF BOND

1 Is appropriate to an individual,

BY THIS BOND [1]We . . . .

of . . . . in the

2 to a Limited Company and

County of . . . . [2]We . . . . Limited

whose registered office is at . . . . in the

3 to a Firm. Strike out whichever two are inappropriate.

County of . . . . [3]We . . . .

and . . . . carrying on business in partnership under

the name or style of . . . .

at . . . . in the

4 Is appropriate where there are two individual Sureties,

County of . . . . (hereinafter called 'the Contractor') [4]and . . . .

. . . . of . . . .

in the County of . . . . and . . . .

of . . . . in the County of . . . .

5 where the Surety is a Bank or Insurance Company. Strike out whichever is inappropriate.

. . . . [5]and . . . . Limited

whose registered office is at . . . . in the

County of . . . . (hereinafter called 'the [4]Sureties/Surety') are held firmly

bound unto . . . . (hereinafter

called 'the Employer') in the sum of . . . . pounds

(£ . . . .) for the payment of which sum the Contractor and the [4]Sureties/Surety bind

themselves their successors and assigns jointly and severally by these presents.

Sealed with our respective seals and dated this . . . . day of . . . .

19 . . . . . .

WHEREAS the Contractor by an Agreement made between the Employer of the one part and the Contractor of the other part has entered into a Contract (hereinafter called 'the said Contract') for the Investigation as therein mentioned in conformity with the provisions of the said Contract.

NOW THE CONDITIONS of the above-written Bond are such that if:

(a) the Contractor shall subject to Condition (c) hereof duly perform and observe all the terms provisions conditions and stipulations of the said Contract on the Contractor's part to be performed and observed according to the true purport intent and meaning thereof or if

(b) on default by the Contractor the Sureties/Surety shall satisfy and discharge the damages sustained by the Employer thereby up to the amount of the above-written Bond or if

(c) the Engineer named in Clause 1 of the said Contract shall pursuant to the provisions of Clause 61 thereof issue an Acceptance Certificate then upon the date stated therein (hereinafter called 'the Relevant Date')

this obligation shall be null and void but otherwise shall remain in full force and effect but no alteration in the terms of the said Contract made by agreement between the Employer and the Contractor or in the extent or nature of the Investigation to be carried out completed and maintained thereunder and no allowance of time by the Employer or the Engineer under the said Contract nor any forbearance or forgiveness in or in respect of any matter or thing concerning the said Contract on the part of the Employer or the said Engineer shall in any way release the Sureties/Surety from any liability under the above-written Bond.

160

PROVIDED ALWAYS that if any dispute or difference shall arise between the Employer and the Contractor concerning the Relevant Date or otherwise as to the withholding of the Acceptance Certificate then for the purposes of this Bond only and without prejudice to the resolution or determination pursuant to the provisions of the said Contract of any dispute or difference whatsoever between the Employer and Contractor the Relevant Date shall be such as may be:

(a) agreed in writing between the Employer and the Contractor or
(b) if either the Employer or the Contractor shall be aggrieved at the date stated in the said Acceptance Certificate or otherwise as to the issue or withholding of the said Acceptance Certificate the party so aggrieved shall forthwith by notice in writing to the other refer any such dispute or difference to the arbitration of a person to be agreed upon between the parties or (if the parties fail to appoint an arbitrator within one calendar month of the service of the notice as aforesaid) a person to be appointed on the application of either party by the President for the time being of the Institution of Civil Engineers and such arbitrator shall forthwith and with all due expedition enter upon the reference and make an award thereon which award shall be final and conclusive to determine the Relevant Date for the purposes of this Bond. If the arbitrator declines the appointment or after appointment is removed by order of a competent court or is incapable of acting or dies and the parties do not within one calendar month of the vacancy arising fill the vacancy then the President for the time being of the Institution of Civil Engineers may on the application of either party appoint an arbitrator to fill the vacancy. In any case where the President for the time being of the Institution of Civil Engineers is not able to exercise the aforesaid functions conferred upon him the said functions may be exercised on his behalf by a Vice-President for the time being of the said Institution.

Signed Sealed and Delivered by the said
  in the presence of:

The Common Seal of

LIMITED

was hereunto affixed in the presence of:

(*Similar forms of Attestation Clause for the
  Sureties or Surety*)

The Institution of
Civil Engineers

The Association of
Consulting Engineers

The Federation of Civil
Engineering Contractors

This clause has been prepared by the Institution of Civil Engineers, the Association of Consulting Engineers and the Federation of Civil Engineering Contractors, in consultation with the Government in its revised form, for use in appropriate cases as a Special Condition of the Conditions of Contract for use in connection with GROUND INVESTIGATION 1983

CONTRACT PRICE FLUCTUATIONS

(1)   The amount payable by the Employer to the Contractor upon the issue by the Engineer of an interim certificate pursuant to Clause 60(2) or of the final certificate pursuant to Clause 60(3) (other than amounts due under this Clause) shall be increased or decreased in accordance with the provisions of this Clause if there shall be any changes in the following Index Figures compiled by the Department of the Environment and published by Her Majesty's Stationery Office (HMSO) in the Monthly Bulletin of Construction Indices (Civil Engineering Works):

    (a)   the Index of the Cost of Labour in Civil Engineering Construction;

    (b)   the Index of the Cost of Providing and Maintaining Constructional Plant and Equipment;

    (c)   the Indices of Constructional Material Prices applicable to those materials listed in sub-clause (4) of this Clause.

The net total of such increases and decreases shall be given effect to in determining the Contract Price.

(2)   For the purpose of this Clause:

    (a)   'Final Index Figure' shall mean any Index Figure appropriate to sub-clause (1) of this Clause not qualified in the said Bulletin as provisional;

    (b)   'Base Index Figure' shall mean the appropriate Final Index Figure applicable to the date 42 days prior to the date for the return of tenders;

    (c)   'Current Index Figure' shall mean the appropriate Final Index Figure to be applied in respect of any certificate issued or due to be issued by the Engineer pursuant to Clause 60 and shall be the appropriate Final Index Figure applicable to the date 42 days prior to:

        (i)   the due date (or extended date) for completion; or

        (ii)   the date certified pursuant to Clause 48 of completion of the whole of the Investigation; or

        (iii)   the last day of the period to which the certificate relates;

    whichever is the earliest.

    Provided that in respect of any work the value of which is included in any such certificate and which work forms part of a Section for which the due date (or extended date) for completion or the date certified pursuant to Clause 48 of completion of such Section precedes the last day of the period to which the certificate relates the Current Index Figure shall be the Final Index Figure applicable to the date 42 days prior to whichever of these dates is the earliest.

    (d)   The 'Effective Value' in respect of the whole or any Section of the Investigation shall be the difference between:

        (i)   the amount which in the opinion of the Engineer is due to the Contractor under Clause 60(2) (before deducting retention) or the amount due to the Contractor under Clause 60(3) (but in each case before deducting sums previously paid on account) less any amounts for Dayworks Nominated Sub-contractors or any other items based

162

on actual cost or current prices and any sums for increases or decreases in the Contract Price under this Clause;

and:

(ii) the amount calculated in accordance with (i) above and included in the last preceding interim certificate issued by the Engineer in accordance with Clause 60.

Provided that in the case of the first certificate the Effective Value shall be the amount calculated in accordance with sub-paragraph (i) above.

(3)   The increase or decrease in the amounts otherwise payable under Clause 60 pursuant to sub-clause (1) of this Clause shall be calculated by multiplying the Effective Value by a Price Fluctuation Factor which shall be the net sum of the products obtained by multiplying each of the proportions given in (a) (b) and (c) of sub-clause (4) of this Clause by a fraction the numerator of which is the relevant Current Index Figure minus the relevant Base Index Figure and the denominator of which is the relevant Base Index Figure.

(4)   For the purpose of calculating the Price Fluctuation Factor the proportions referred to in sub-clause (3) of this Clause shall (irrespective of the actual constituents of the work) be as follows and the total of such proportions shall amount to unity:

  (a)   0.___* in respect of labour and supervision costs subject to adjustment by reference to the Index referred to in sub-clause (1)(a) of this Clause;
  (b)   0.___* in respect of costs of provision and use of all civil engineering plant road vehicles etc. which shall be subject to adjustment by reference to the Index referred to in sub-clause (1)(b) of this Clause:
  (c)   the following proportions in respect of the materials named which shall be subject to adjustment by reference to the relevant indices referred to in sub-clause (1)(c) of this Clause:

  0.___* in respect of Aggregates

  0.___* in respect of Bricks and Clay Products generally

  0.___* in respect of Cements

  0.___* in respect of Cast Iron products

  0.___* in respect of Coated Roadstone for road pavements and bituminous products generally

  0.___* in respect of Fuel for plant to which the Gas Oil Index will be applied

  0.___* in respect of Timber generally

  0.___* in respect of Reinforcing steel (cut, bent and delivered) and other metal sections

  (d)   0.10___ in respect of all other costs which shall not be subject to any adjustment;

Total   1.00 .

(5)   Provisional Index Figures in the Bulletin referred to in sub-clause (1) of this Clause may be used for the provisional adjustment of interim valuations but such adjustments shall be subsequently recalculated on the basis of the corresponding Final Index Figures.

(6)   Clause 69—Tax Fluctuations—shall not apply except to the extent that any matter dealt with therein is not covered by the Index of the Cost of Labour in Civil Engineering Construction.

* To be filled in by the Employer prior to inviting tenders.

163

Appendix 2
Schedules of Dayworks Carried out Incidental to Ground Investigation Site Operations

1. Specialist and non-specialist labour
2. Materials
3. Specialist plant with crew
4. Specialist plant and equipment without crew
5. Supplementary charges
6. Value Added Tax

These Schedules are the Schedules referred to in Clause 52(3) of the ICE Conditions of Contract for Ground Investigation and have been prepared for use in connection with Dayworks carried out incidental to contract work where no other rates have been agreed. They are not intended to be applicable for Dayworks ordered to be carried out after the site operations have been substantially completed or to a contract to be carried out wholly on a daywork basis. The circumstances of such works vary so widely that the rates applicable call for special consideration and agreement between Contractor and Engineer.

1 Specialist and non-specialist labour
Preamble
The rates given in the specialist and non-specialist labour schedules are to be all inclusive not subject to any percentage additions and therefore are to include all statutory charges applicable at the date for return of tenders and all other charges and costs including:

Wages, bonus, daily travelling allowances (fare and/or time), tool allowance and all prescribed payments including those in respect of time lost due to inclement weather paid to workmen at plain time rates and/or at overtime rates.

Subsistence or lodging allowances and periodic travel allowances (fare and/or time) paid to or incurred on behalf of workmen.

Standard rate national insurances.

Normal contract works, third party and employer's liability insurances.

Annual and public holidays with pay.

Non–contributory sick pay scheme.

Industrial training levy.

Redundancy payments contribution.

Obligations under Contracts of Employment Act.

Site supervision and staff.

Small tools – such as picks, shovels, barrows, trowels, hand saws, buckets, trestles, hammers, chisels and all items of a like nature.

Protective clothing.

Head office charges and profit.

1.1 Specialist schedule

Item no.	Description	Period	Hire rate (£)
1.1.1	Investigation supervisor	Per hour	
1.1.2	Rotary drilling foreman operator	Per hour	
1.1.3	Rotary drilling assistant	Per hour	
1.1.4	Rotary drilling hand	Per hour	
1.1.5	Boring rig foreman operator	Per hour	
1.1.6	Boring rig assistant	Per hour	
1.1.7	Boring rig hand	Per hour	
1.1.8	Senior technician	Per hour	
1.1.9	Technician	Per hour	
1.1.10	Assistant technician	Per hour	

The Engineer should add to the schedule any additional categories for which he requires Daywork rates to be entered, and delete any which he deems to be inappropriate.

1.2 Non-specialist schedule

Item no.	Description	Period	Hire rate (£)
1.2.1	Fitter	Per hour	
1.2.2	Welder	Per hour	
1.2.3	Electrician	Per hour	
1.2.4	Scaffolder	Per hour	
1.2.5	Driver	Per hour	
1.2.6	Watchman	Per hour	
1.2.7	Labourer	Per hour	
1.2.8	Timberman foreman	Per hour	
1.2.9	Timberman	Per hour	
1.2.10	Crane driver	Per hour	
1.2.11	Banksman	Per hour	
1.2.12	Digger driver	Per hour	
1.2.13	Boat man	Per hour	

2 Materials
Preamble

The Contractor shall state the percentage to be added to the cost of materials delivered to site. This percentage shall allow for the following.

The percentage addition provides for Head Office charges and profit.

The cost of materials means the invoiced price of materials including delivery to site without deduction of any cash discounts not exceeding 2½ per cent.

Unloading of materials: the percentage added to the cost of the materials includes the cost of handling.

Note: if the Contractor wishes to place a different percentage on certain materials he shall give details.

2.1 Add to the cost of materials delivered to site .... %

3 Specialist plant with crew
Preamble

The rates given below are to include for the charges as set out in the preamble to Schedule 1 for each and every member of the crew.

The rates apply only to specialist plant already on site and are to include: consumable stores, repairs, maintenance, insurance, fuel including distribution and general servicing, and are to be paid for in whole units of time for each item and for each period of hire.

3.1 Schedule for specialist plant with crew

Item no.	Description	Unit	Hire rate (£)	Period
3.1.1	Rotary drill and crew (quote types and capacity for each rig) including pump and all necessary drilling and coring equipment.	Each		Per hour
3.1.2	Extra over 3.1.1 for flushing systems other than water (state type).	Each		Per hour
3.1.3	Extra over 3.1.1 for special equipment (state type).	Each		Per hour
3.1.4	Light cable percussion boring rig with crew (quote types and capacity for each rig) and all necessary boring equipment, casing and SPT equipment and U100 sampling gear.	Each		Per hour
3.1.5	Extra over 3.1.4 for rotary drilling attachment including pumps and all necessary drilling and casing equipment (state type).	Each		Per hour
3.1.6	Extra over 3.1.4 for casing oscillator (state type).	Each		Per hour

The Engineer should add to the schedule any additional specialist plant and equipment for which he requires Daywork rates to be entered and delete any which he deems to be inappropriate.

4 Specialist plant and equipment without crew
Preamble

The rates apply only to specialist plant already on site and are to include: consumable stores, repairs, maintenance, insurance, and fuel including distribution, general servicing and are to be paid for in whole units of time for each item and for each period of hire.

Hire rates for other special plant not normally classed as small tools and not included in this section shall be fixed at prices reasonably related to the rates quoted.

4.1 Schedule for specialist plant and equipment

Item no.	Description	Unit	Hire rate (£)	Period
4.1.1	Additional pump (state type).	Each		Per hour
4.1.2	Static cone penetration test equipment (state type and capacity).	Each		Per hour
4.1.3	Site vane test equipment (state type).	Each		Per hour
4.1.4	Specialist sampling equipment (state type).	Each		Per hour
4.1.5	Specialist test equipment (state type).	Each		Per hour

The Engineer should add to the schedule any additional specialist plant and equipment for which he requires Daywork rates to be entered and delete any which he deems to be inappropriate.

5 Supplementary charges
Preamble

The cost of transporting to site of additional crew, plant and equipment for agreed Dayworks additional to that already on site shall be charged at cost plus the percentage stated below.

The cost of additional insurance premiums for abnormal contract work or special site conditions to be charged at cost plus the percentage stated below.

5.1   *Add to the cost of providing the transportation referred to in the first paragraph above* .... %

5.2   *Add to the cost of the additional insurance premiums referred to in the second paragraph above* .... %

6 Value Added Tax

Value Added Tax has not been included in any of the rates in this schedule but will be chargeable if payable to HM Customs and Excise by the Contractor.

169

Appendix 3
Proposed Schedules of Dayworks Carried out Incidental to Ground Investigation Site Operations, presented for consideration by F.H. Hughes and H.G. Clapham

G.1. Specialist and non-specialist labour
G.2. Materials
G.3. Specialist plant with crew
G.4. Specialist plant and equipment without crew
G.5. Supplementary charges
G.6. Value Added Tax

G.1 Specialist and non-specialist labour
Preamble
The rates given in the Specialist and Non-Specialist Labour Schedules are to be all inclusive not subject to any percentage additions and therefore are to include all statutory charges applicable at the date for return of tenders and all other charges and costs including:

Wages, bonus, daily travelling allowances (fare and/or time), tool allowance and all prescribed payments including those in respect of time lost due to inclement weather paid to workmen at plain time rates and/or at overtime rates.
Subsistence or lodging allowances and periodic travel allowances (fare and/or time) paid to or incurred on behalf of workmen.
Standard rate national insurances.
Normal contract works, third party and employers' liability insurances.
Annual and public holidays with pay.
Non-contributory sick pay scheme.
Industrial training levy.

170

Redundancy payments contribution.
Obligations under Contracts of Employment Act.
Site supervision and staff.
Small tools – such as picks, shovels, barrows, trowels, hand saws, buckets, trestles, hammers, chisels and all items of a like nature.
Protective clothing.
Head office charges and profit.

G.1.1 Specialist schedule

Item no.	Description	Period	Hire rate (£)
1.1.1	Investigation Supervisor	Per hour	
1.1.2	Rotary drilling foreman operator	Per hour	
1.1.3	Rotary drilling assistant	Per hour	
1.1.4	Rotary drilling hand	Per hour	
1.1.5	Boring rig foreman operator	Per hour	
1.1.6	Boring rig assistant	Per hour	
1.1.7	Boring rig hand	Per hour	
1.1.8	Senior technician	Per hour	
1.1.9	Technician	Per hour	
1.1.10	Assistant technician	Per hour	

G.1.2 Non-specialist schedule

Item no.	Description	Period	Hire rate (£)
1.2.1	Fitter	Per hour	
1.2.2	Welder	Per hour	
1.2.3	Electrician	Per hour	
1.2.4	Scaffolder	Per hour	
1.2.5	Driver	Per hour	
1.2.6	Watchman	Per hour	
1.2.7	Labourer	Per hour	
1.2.8	Timberman foreman	Per hour	
1.2.9	Timberman	Per hour	
1.2.10	Crane driver	Per hour	
1.2.11	Banksman	Per hour	
1.2.12	Digger driver	Per hour	
1.2.13	Boat man	Per hour	

The Engineer should add to the schedule any additional categories for which he requires Daywork rates to be entered, and delete any which he deems to be inappropriate.

G.2 Materials
Preamble
The Contractor shall state the percentage to be added to the cost of materials delivered to site. This percentage shall allow for the following.

The percentage addition provides for Head Office charges and profit.
The cost of materials means the invoiced price of materials including delivery to site without deduction of any cash discounts not exceeding 2½ per cent.
Unloading of materials: the percentage added to the cost of the materials includes the cost of handling.
Note: if the Contractor wishes to place a different percentage on certain materials he shall give details.

G.2.1
Add to the cost of materials delivered to site......................%

G.3 Specialist plant with crew
Preamble
The rates given below are to include for the charges as set out in the preamble to Schedule 1 for each and every member of the crew.

The rates apply only to specialist plant already on site and are to include: consumable stores, repairs, maintenance, insurance, fuel including distribution and general servicing, and are to be paid for in whole units of time for each item and for each period of hire.

The Engineer should add to the schedule any additional specialist plant and equipment for which he requires Daywork rates to be entered and delete any which he deems to be inappropriate.

G.3.1 Schedule for specialist plant with crew

Item no.	Description	Unit	Hire rate (£)	Period
G.3.1.1	Rotary drill and crew (quote types and capacity for each rig) including pump and all necessary drilling and coring equipment.	Each		Per hour
G.3.1.1.1	Working.	Each		Per hour
G.3.1.1.2	Standing.	Each		Per hour
G.3.1.2	Extra over 3.1.1.1 or 3.1.1.2 for flushing systems other than water (state type).	Each		Per hour
G.3.1.3	Extra over 3.1.1.1 or 3.1.1.2 for special equipment (state type).	Each		Per hour
G.3.1.4	Light cable percussion boring rig with crew (quote types and capacity for each rig) and all necessary boring equipment, casing and SPT equipment and U100 sampling gear.	Each		Per hour
G.3.1.4.1	Working.	Each		Per hour
G.3.1.4.2	Standing.	Each		Per hour
G.3.1.5.1	Extra over 3.1.4.1 for rotary drilling attachment including pumps and all necessary drilling and casing equipment (state type).	Each		Per hour
G.3.1.5.2	Extra over 3.1.4.2 for rotary drilling attachment including pumps and all necessary drilling and casing equipment (state type).	Each		Per hour
G.3.1.6	Extra over 4.1.4.1 or 3.1.4.2 for casing oscillator (state type).	Each		Per hour

G.4 Specialist plant and equipment without crew
Preamble
The rates apply only to specialist plant already on site and are to include: consumable stores, repairs, maintenance, insurance, and fuel including distribution, general servicing and are to be paid for in whole units of time for each item and for each period of hire.

Hire rates for other special plant not normally classed as small tools and not included in this section shall be fixed at prices reasonably related to the rates quoted.

G.4.1 Schedule for specialist plant and equipment working

Item no.	Description	Unit	Hire rate (£)	Period
G.4.1.1	Additional pump (state type).	Each		Per hour
G.4.1.2	Static cone penetration test equipment (state type and capacity).	Each		Per day
G.4.1.3	Site vane test equipment (state type).	Each		Per day
G.4.1.4	Specialist sampling equipment (state type).	Each		Per day
G.4.1.5	Specialist test equipment (state type).	Each		Per day

The Engineer should add to the schedule any additional specialist plant and equipment for which he requires Daywork rates to be entered and delete any which he deems to be inappropriate.

(Schedules dealing with Supplementary Charges and Value Added Tax follow. Eventually, no doubt, there will be an additional schedule covering the processes, apparatus and protective clothing applicable to the investigation of contaminated land.)

Schedule G.5
Schedule 4 of the FCED document covers, where applicable, the cost of free transport and the use of sub–contractors where necessary. Schedule G.5 is designed for mobilisation of extra resources.

In some cases it is necessary in Ground Investigation to cover a hazard beyond normal contract risks. An example might be where areas of unstable ground are to be investigated or marine works to be undertaken.

G.5 Supplementary Charges
Preamble
The cost of transporting to site, and removal on completion of additional crew, plant and equipment required for agreed Dayworks additional to that already on site shall be charged at cost plus the percentage stated below.

G.5.1.
Add to the cost of providing the transportation referred to in the paragraph above...%

The cost of additional insurance premiums for abnormal contract work or special site conditions to be charged at cost plus the percentage stated.

G.5.2.
Add to the cost of the additional insurance premiums referred to in the paragraph above ....%

Schedule G.6
This is identical to the clause given in the FCEC document.

G.6 Value Added Tax
Value Added Tax has not been included in any of the rates in this schedule but will be chargeable if payable to HM Customs and Excise by the Contractor.

British Standards referred to in the text

BS 812: *Methods for sampling and testing mineral aggregates, sands and fillers.*
Part 1: 1975 Sampling, size, shape and classification.
Part 2: 1975 Physical properties.
Part 3: 1975 Mechanical properties.
Part 4: 1976 Chemical properties.

BS 1377: *1975 Methods for test for soil for civil engineering purposes.*

BS 1924: *1975 Methods of test for stabilized soils.*

BS 2091: *1969 Respirators for protection against harmful dust, gases and scheduled agricultural chemicals.*

BS 3776: *1964 Rescue lines for industrial workers.*

BS 5573: *1978 Code of practice for safety precautions in the construction of large diameter boreholes for piling and other purposes (formerly CP 2011).*

BS 5930: *1981 Code of practice on site investigation.*

BS 0000: *(unallocated) Code of practice for the identification and investigation of contaminated land.*

BS 6031: *1981 Code of practice for earthworks.*

CP 2001: *1957 Site investigations.*

CP 2003: *1959 Earthworks.*

CP 2004: *1972 Foundations.*

Index